Genesis and Evolutionary Development of Life

Genesis and Evolutionary Development of Life

A. I. OPARIN
Active Member of the Academy of Sciences of the U.S.S.R.

TRANSLATED FROM THE RUSSIAN BY ELEANOR MAASS

ACADEMIC PRESS New York and London **1968**

First published in the Russian language under the title *Vozniknoveniye i nachal'noye razvitiye ahizni* by MEDITSINA PUBLISHING HOUSE, Moscow, U.S.S.R. in 1966.

ACADEMIC PRESS, INC.
111 Fifth Avenue, New York, New York 10003

United Kingdom Edition published by
ACADEMIC PRESS, INC. (LONDON) LTD.
Berkeley Square House, London W.1

LIBRARY OF CONGRESS CATALOG CARD NUMBER: 69-12278

PRINTED IN THE UNITED STATES OF AMERICA

PREFACE TO THE ENGLISH EDITION

Publication of the Darwinian theory of evolution is justly considered the beginning of a new era in biology — when the science of life was elevated to a previously unattainable height. Before Darwin, biology was an accumulation of scattered facts which, although systematized, was not intrinsically interrelated. The tremendous significance of Darwin's doctrine was that it subordinated all of this material to a single idea of development; it imparted an intrinsic meaning not only to previously investigated biological material, but also to newly discovered material.

Now, however, Darwinism appears to us as a "glittering iceberg," almost nine-tenths of which is hidden from view under water. Prebiological and biological evolution, in the process of which basic features and qualities characteristic of all life were formed, took place considerably before the origin of objects usually studied by Darwinists.

As recently as the first half of this century, knowledge of that hidden segment of evolution was completely unattainable to man. Attempts were made to explain the advent of the forebears of all life on Earth, that is, forebears according to Darwinian evolution, either by the transport of completely organized microorganisms to our planet from other worlds or by the sudden emergence of such organisms as the result of some sort of unique, extremely rare "lucky accident." In both cases, the evolutionary pathway by which the first living beings emerged was completely ignored or even denied.

At the present time, a fundamental change has occurred in the thinking of different groups of naturalists regarding the problem which interests us. Today, the conclusion has become generally accepted that the origin of life was by no means a "lucky accident," but a phenomenon completely approachable by objective scientific study.

The abundance of data accumulated by natural scientists at the present time has made it possible, to a certain degree, to imagine the sytematic progress of this development of ours on Earth. This material

v

has opened wide the door to investigations striving not only to explain theoretically the greatest events of the past, but also to obtain experimental proof of the accuracy of these explanations.

The rapid growth of this type of investigation, in which scientists of various disciplines in many parts of the world are participating, requires a broad exchange of ideas and, possibly, a complete review of the results obtained. This is necessary because "the origin of life" has now attained a widespread irresistible fascination for all of mankind — every thinking individual is interested in it. However, to accomplish this feat is certainly not easy in view of the very rapid growth of our knowledge in the areas which interest us.

I am very grateful to Academic Press for undertaking the task of publishing the English translation of my book in which I have attempted, in as popular a form as possible, to make known to a wide circle of readers the present state of thought on the origin and development of life.

September, 1968   A. I. OPARIN

CONTENTS

Genesis and Evolutionary
Development of Life

INTRODUCTION

Observing nature around us, we readily divide it into the nonliving, inorganic world, and the world of living beings. The world of living beings is represented by an enormous variety of animals, plants, and microorganisms. Externally they are very different, and yet we immediately notice in them something in common which distinguishes them from objects of the inorganic world and that compels us to include in the single concept "living being" dog and tree, whale and insignificant insect or blade of grass, man and bacterium barely distinguishable under the microscope. There is something common inherent in all of these that we call life.

But what is the essence of this "something"? What is the essence of life? Without a scientifically based solution to this question, it is impossible to construct a view of the world correctly reflecting objective reality.

Furthermore, this solution affects the practical activity of man in all those areas where man deals with living beings, for example in agriculture or in medicine. In all these cases the influence of man on living nature will be more fruitful, the deeper he penetrates into the essence of life, the better he understands it.

History teaches us that during the whole conscious life of mankind, the problem of the essence of life has been one of the basic battlefields in the bitter ideological struggle of two irreconcilable philosophical camps — idealism and materialism. Members of the idealistic camp view the essence of life as having some sort of eternal supernatural source incomprehensible by the experimental method. This is the "psyche" of Plato, the "entelechy" of Aristotle, the immortal soul or divine spark of various religious scholars and faiths, the Kantian "inner principle of action," the manifestation of the Hegelian "universal reason," the life force" of the vitalists, the dominating idea of the neovitalists and so forth.

Matter, that objective reality which we comprehend broadly and fully

1

by direct observation and experimental study, from the idealistic position is nonliving and inert. It serves only as material from which the soul or mind creates living essence.

Proceeding from the idealistic concepts, it is practically impossible to discover the essence of life by the method of objective study of the nature surrounding us on materialistic bases since life itself is supernatural and spiritual. Only by means of a deepened, speculative self-knowledge can we supposedly approach an understanding of that divine origin which we carry within ourselves. The whole remaining world of living beings we can only passively contemplate and yet, of course, there can be no question, from this point of view, of any variation or any alteration of living nature by man.

Materialism approaches the problem of the essence of life from a diametrically opposite point of view; materialism, based on facts derived from natural science, confirms that life, like all the rest of the world, is material in its nature, and it does not need for its understanding any sort of spiritual origin incomprehensible by experimental means. On the contrary, from the materialistic viewpoint it is precisely the objective study of the nature surrounding us which is the reliable means not only leading us to a perception of the essence of life itself, but also permitting us directly to vary or improve living nature for the welfare of man. Many groups of biologists and naturalists consciously or unconsciously begin their investigative work from a materialistic perception of living nature. Proceeding in this way, they increasingly enrich by their work the science of life, bringing us closer to an understanding of its essence. However, within the limits of this materialistic concept of life its essence can be comprehended in many different ways.

Many scientists of the last century and modern investigators, basing their attitudes on a mechanistic understanding of life, considered and still consider that knowledge of life in general consists only in the most complete elucidation of it by physics and chemistry, in the most complete reduction of all living phenomena to physical and chemical processes. From these viewpoints, there are no specifically biological laws. There actually exists one law alone ruling inorganic nature, which also directs all phenomena occurring in living bodies. Thus mechanistic materialism denies any sort of qualitative difference between organisms and objects of inorganic nature.

In contrast to this, dialectical materialism regards life as a special form of the motion of matter, qualitatively differentiated from objects of the inorganic world; this form of motion is inherent only in living beings.

According to dialectical materialism, matter is in constant motion; it is not something external to matter, but is necessarily connected with the essence of matter itself. The motion of matter in this sense must not be regarded merely as the transposition of material bodies in space. Such mechanical transposition is only one (and furthermore, the lowest, simplest) form of the motion of matter. In addition, other more complex forms of motion are also inherent in a variety of objects of activity; these forms arise as a new quality in the process of development of matter.

This process of development we can observe everywhere in the world around us. It is progressive in nature, and is directed to the creation of higher and higher forms of matter in motion. The new quality of material objects arising in the process of development is completely real, but they were not preceded before their creation by some latent form; they actually appear afresh only at a given stage of the development of matter, as more complex, diverse, and abundantly differentiated forms of motion.

Furthermore, as new qualities arise, the previous forms of matter in motion do not disappear. They, too, are preserved in the qualitatively new objects of tangible reality, but fade into the background, as it were. They have relatively little significance for the subsequent development of these objects in comparison with the significance of new forms of motion. The further fate of the qualitatively new natural bodies is determined in general still by their special properties and the new objective relationships, which it is impossible to reduce by means of simple analysis to the previous more primitive forms. Every representation of this sort of reducibility is equivalent in essence to a denial of genuine development, it distorts reality, and is inconsistent with true dialectical nature. But this is just what the mechanists do in respect to their comprehension of life. Completely ignoring the process of development of matter, they strive to comprehend life purely metaphysically. Their method, roughly speaking, consists in dismantling the organism, like a clock, into separate little screws and wheels, and on the basis of a study of these parts, comprehending the living being as a whole.

Of course, there is no doubt that detailed analysis of substances and phenomena characteristic of a living body is extremely important, entirely necessary for the correct understanding of life. The brilliant successes of modern biochemists and biophysicists is graphic proof of this. Studying in detail only the chemistry of the material substrate of life, studying individual substances and their composition, individual reactions taking part in metabolism and their linkages, recognizing the bioenergetic structural and general physical properties of life, we can be rationally directed toward a scientific comprehension of life. To prove the prominent role of physical and chemical investigations in this respect is like breaking down an unlocked door. It is obvious, but the whole question consists in whether these investigations alone are sufficient in themselves to attain the set goal, to comprehend the essence of life. Evidently, this is not so.

Dialectical materialism, regarding life as a qualitatively special form of the motion of matter, defines the task itself of comprehension of life in a different way than mechanistic materialism. Mechanistically, the problem consists in the most complete reduction of living phenomena to physical and chemical processes. On the contrary, from a dialectical materialistic point of view, comprehension of life consists mainly in the establishment of its precise qualitative distinctions from other forms of the motion of matter. The clearest expression of life (as a special form of the motion of matter) is found in the specific interaction of living systems — organisms — with the environment surrounding them, in dialectical unity of the living body and the conditions of its existence.

At the basis of this unity lies biological metabolism. In metabolism many tens and perhaps hundreds of thousands of individual biochemical reactions are joined to each other in orderly fashion in a complex network* of transformations of matter and energy. Closely bound in this network are not only the processes of synthesis, assimilation, as a result of which the substances of the environment are converted into protein and other compounds characteristic of organisms, but also the processes of dissimilation, breakdown, mainly serving as a source of the energy necessary for life. As a result of this, any organism during its whole life is in continual internal motion, in a state of continual breakdown and synthesis. Its visible stability is only an external expression of the exceedingly efficient coordination of these opposing branches of metabolism, as a result of which a newly formed particle replaces each decomposing particle of protein or other substance. In this

---

*The term "network" is understood here and later on, not in a spatial, but in a time sense and refers to the coupling of metabolic reactions.

way the organism preserves to a certain degree the constancy of its form and chemical composition, always being changed materially thereby. A characteristic qualitatively distinguishing life as a higher form of the motion of matter is the fact that in living bodies, numerous biochemical reactions, consisting of metabolism in its aggregate, not only are strictly coordinated among each other in time and in space, not only are united into a single sequence of continuous self-renewal, but also the whole of this order is directed toward constant self-preservation and self-reproduction of the entire living system, is exceedingly efficiently adapted to solution of the task of survival of the organism in the given environmental conditions.

This efficient adaptation, or as it is often called, "purposiveness" of organization runs through the whole living world from top to bottom, to the most elementary forms of life. It is a necessary property of any living body, but apart from it, in the natural state of inorganic nature, this purposiveness is absent. It is specific for life as a quality, formed as something new in the process of the establishment of life. Thus non-living inorganic bodies lack this "purposiveness" and we would vainly look for its explanation solely in the relationships of the inorganic world, solely in the laws of physics and chemistry.

Idealists see in this "purposiveness" proof of the supernatural nature of life – the purposeful fulfillment by living beings of the plan of some sort of higher spiritual source. The only thing the mechanists can evoke in this connection is the likening of organisms to artificially formed mechanisms, to machines. The construction of a machine, its internal organization, say the mechanists, is always adapted to the fulfillment of certain specific work, it is "purposeful." Apart from life it is only in machines that we find that sort of adaptability of structure to fulfilling functions, and in addition to this, the work of any machine can be as a whole reduced to physical and chemical phenomena. On this basis, even as early as Descartes and up to our day tendencies exist to regard an organism as some sort of very complex mechanism. In a different epoch only the material appearance of this tendency has been changed in connection with the development of science and technology. Thus, in the seventeenth century living beings were compared with tower clocks, later with steam engines, and now with cybernetic machines, capable of fulfilling a series of functions in the intellectual labor of man. This is why cybernetics is often regarded now as an entirely new, universal way of getting to know the essence of life itself.

This, of course, is false. Attempts to regard living essence as a sort of mechanism have existed for many centuries. However, the more deeply we study the organization of living beings, the more nominal becomes the analogy between organisms and mechanisms. No matter how complex and efficient an organization the modern electronic computer has, it is farther removed in its nature from man, than, for example, the most primitive bacteria, although the latter also does not possess that differentiated nervous system which the machine so successfully imitates.

This difference between organisms and mechanisms is distinctly manifested also in their material nature, in the fact that living beings, according to the expression of Engels, are "proteinaceous bodies," and in the continuous character of the organization of metabolism, and in the character of the energetics of living nature. It is especially important to note here that the analogy advanced by mechanists between organisms and machines cannot in any way explain precisely that which it should explain – the "purposiveness" of the organization of living beings. But machines are not inorganic systems simply acting on the basis of physical and chemical relationships alone. They are the offspring not of biological, but of a higher social form of the motion of matter – the social life of humans. The inner "purposiveness" of the machine, the adaptability of its structure to the fulfillment of certain work, cannot be revealed from the interaction of some law of the inorganic world, it is the fruit of the psychic activity of man, his creative efforts. Thus, when mechanists attempt to explain the inner adaptability of organisms by analogy to machines, they inevitably arrive at an especially idealistic conclusion – at the recognition that "purposiveness" of the organization of life is the result of the creative wish of a creator, "godly quantum mechanics," or simply speaking, at a recognition of the divine origin of life.

Thus mechanistic materialism, outwardly in opposition to idealism, steps into the same blind alley as the latter in its attempts to comprehend the essence of life, precisely because it metaphysically ignores the dialectical route of the development of matter, it strives to comprehend life apart from its formation.

Rationally, life can be materialistically comprehended only as a special form of the motion of matter, arising in a regular manner at a certain stage in the development of matter.

The great dialectician, the ancient Greek Heraclitus of Ephesus, and after him Aristotle, taught that we can only understand the essence of things when we know their origin and development. These words full of deep meaning, of course, are true as a whole in respect to the comprehension of life.

Just by studying the route of its origin, we can understand how and why just these, and no other features characteristic of life were fashioned, how in this process new qualities and biological relationships formerly lacking arose, and how that purposiveness of organization was formulated which so strikes us in all living beings.

Thus modern naturalists have become more and more consciously convinced that the problem of the essence of life and the problem of its origin are inseparable, are only two sides of the same coin. In addition to this it has now become more and more evident that it is impossible to regard (as had been the case until recently) the emergence of life as some sort of sudden isolated phenomenon, as some sort of "lucky accident." It is an inherent component part of the overall process of development of the universe, proceeding in an orderly fashion.

The emergence of our terrestrial life is only an individual, special branch of this overall development. The special feature of this branch is that at its base lies the ordered evolution of carbon compounds and the complex systems formed from them.

The beginning stages of this evolution are very universal, widely dispersed in the cosmos. They can be discovered in quite diverse heavenly bodies. However, the later stages in the development of carbon formations were specific for terrestrial conditions. They were inseparably connected with the evolution of our planet itself.

The large amount of factual material accumulated up to the present time in natural science has made it possible to represent concretely the course of this development on Earth, that threw the doors wide open on objective scientific investigations, attempting not only theoretically to explain the greatest events of the past, but also to obtain experimental proof of the correctness of this explanation. However, natural science did not reach this position at one stroke. Mankind trod a long and tortuous path before the right road was found leading to the resolution of the problem of the origin of life. The history of these searches is very instructive and with it we begin our account.

GENERAL REFERENCES

R. Descartes. "Oeuvres philosophiques." Aime et Marin, Paris, 1839.

F. Engels. "Dialectics of Nature" (transl. by C. Dutt). Foreign Language Press, Moscow, 1954.

Heraclitus of Ephesus. Fragments. English transl. by K. Freeman: De Caelo Ancilla to the Pre-Socratic Philosophers, Oxford Univ. Press, London and New York, 1952.

W. Hollitscher. "Die nature im Weltbild der Wissenschaft." Globus Verlag, Vienna, 1960.

"Istoriya filosofii" ("History of Philosophy"), Vol. 1, p. 464. Izd. Akad. Nauk SSSR, 1940.

P. Kossa. "Kibernetika." Foreign Literature Press, 1958.

A. Oparin. "Life, its Nature, Origin and Development." Academic Press, New York, 1961.

E. Schrodinger. "What is Life?" Cambridge Univ. Press, London and New York, 1945.

Simplicius on Aristotle's De Caelo Ancilla to the Pre-Socratic philosophers. Oxford Univ. Press, London and New York, 1952.

CHAPTER 1

SHORT HISTORY OF ATTEMPTS
TO SOLVE THE PROBLEM OF THE ORIGIN OF LIFE

For many centuries people thought that the Earth was flat and immobile and that the sun revolved around it, rising in the east and setting into the sea or behind the mountains in the west (Fig. 1). This false conviction rested on direct, uncritical observation of the surrounding natural world. The same sort of observation often inspired the thought that various forms of life, for example, insects, worms, and at another time, even fish, birds and mice, not only could be born from things like themselves but also could arise directly by spontaneous generation from slime, manure, soil, and other nonliving materials.

It is clear why these everyday uncritical observations should have had such vast significance in forming the concepts of ancient peoples, since nature had not yet been studied under a microscope, nor subjected to analysis and calculation, but was perceived as a whole in direct contemplation by the senses (1). In China, for example, even from the earliest times, there was belief in the spontaneous generation of aphids and other insects under the influence of heat and moisture (2). In sacred books of India there are indications of the sudden emergence of various parasites, flies, and beetles from sweat and manure (3). It can be deciphered from Babylonian cuneiform texts that worms and other creatures were formed from the mud of canals (4).

In ancient Egypt the conviction prevailed that the layer of humus deposited by the Nile in its flood stage could give birth to living creatures when it was warmed by the sun; frogs, toads, snakes and mice were formed thus. In this case one might easily convince oneself by direct observation that the front part seemed already finished and alive while the hind part was still unchanged damp earth (5). We find a repetition of these tales in the writings of the ancient Greeks (for example, in Diogenes of Appollonia) and in the works of the famous

FIG. 1. Picture of the world, as visualized in the Middle Ages.

Roman scholar Pliny. They were widely disseminated both in the West and in the East, not only in the Middle Ages but also later. No wonder that even in Shakespeare's tragedy "Anthony and Cleopatra" Lepidus asserts that in Egypt the crocodiles emerge from the ooze of the Nile under the influence of the sultry southern sun (6).

In general it is very characteristic of the history of our problem that among the most diverse peoples, living at different times, at different cultural stages, we find indications almost everywhere of spontaneous formation of the same organisms — flies, spawning in manure and rotting meat; lice, forming from human sweat; glowworms, breeding from sparks of a wood fire; finally, there are the frogs and mice emerging from the dew and moist earth. Everywhere that man encountered the sudden and overwhelming appearance of living beings, he conceived of it as the spontaneous generation of life. Faith of ancient peoples in spontaneous generation did not exist because of a conclusion formed about any one concept of the world. To them spontaneous generation of life was simply a tangible, empirically established fact, which later was placed in some sort of theoretical context.

The ancient teachings of India, Babylon, and Egypt regarding the origin of life were based on various religious legends and beliefs. From this point of view spontaneous generation of life was only a particular case of the manifestation of the creative powers of gods and demons. But even at the very source of our European culture, ancient Greece, theogony, as a mythical interpretation of nature, was replaced by cosmogony, as the beginning of scientific investigation (7).

Although all Greek philosophers from the Milesians to Epicurus and the Stoics accepted spontaneous generation of life as an undisputed fact, their philosophical interpretation of this "fact" had already come far from the framework of the previous mystical concept. It contained the beginnings of all those concepts which subsequently were to develop into the problem of the origin of life.

The first of the Greek philosophers, Thales, living about 624–547 B.C. even at this early date approached the essence and origin of life from an elemental materialistic standpoint. Thales, as did other philosophers of the Milesian school (Anaximander and Anaximenes) recognized as a fundamental principle the objective existence of matter as something which is always living, and always changing from the beginning of time. Life is inherent in matter as such. Thus although the Milesians also believed in the spontaneous generation of living beings from slime, mud, etc. they interpreted this phenomenon only as the autoformation of individual organisms, not needing for its accomplishment the intervention of any special mystic force (8). Later Empedocles (485–425 B.C.) (9) developed the point of view that plants and animals were formed from unorganized yet already living substances, either by being bred from "like beings" or from "unlike beings" that is, by spontaneous generation. A particularly clear expression of the idea of spontaneous generation of living beings is found in the teachings of Democritus (460–370 B.C.) (10). Therein ancient Greek materialism reached the summit of its development, although it had already acquired a somewhat mechanistic character.

According to the views of Democritus, the universe is made up of matter consisting of a multitude of extremely small particles (atoms) which are in perpetual motion and are separated by empty space. This mechanical motion of the atoms is inherent in matter and it causes the formation of all individual things. In particular, life does not result from the action of a supernatural creative power, but from the mechanical force of nature itself. According to Democritus, the primordial emergence of life or its spontaneous generation from water and slime occurs as a consequence of an accidental, but completely ordered union of atoms in their mechanical movement, when the almost imperceptible particles of moist earth meet and combine with atoms of fire.

Another distinguished thinker of ancient Greece, Epicurus (342–271 B.C.) (11) took a similar philosophical stand one hundred years later. We find an exposition of his views in the famous poem of Lucretius Carus "On the Nature of Things" (12). According to this source, Epicurus taught that worms and numerous other animals were generated from the soil or manure by the action of the moist warmth of the sun and rain. This occurs, however, without the participation of any sort of spiritual source. The soul as a nonmaterial force, according to Epicurus, does not exist. The soul is material and consists of small, very fine and smooth atoms. The mechanical union of atoms in empty space leads to the genesis of a diversity of things, in particular living beings. The reason for the movement of atoms lies in the matter itself and does not depend on some sort of "initial impulse" or other intervention of the gods into the affairs of the world.

Thus, even a hundred years before our era, the phenomenon of spontaneous generation was explained materialistically by many philosophical schools as a natural process of autoformation of living things without the intervention of any sort of spiritual force. But historically it came about that the later development of the idea of spontaneous generation was linked basically not with the materialistic "Democritian line" but with the idealistic "Platonic line" antagonistic to it.

Plato (427–347 B.C.) hardly concerned himself directly with the problem of spontaneous generation; in "Phaedo" he casually mentions the possibility of the genesis of living things under the influence of heat and putrefaction. However, in complete accord with his general philosophical position, Plato asserted that plants and animal matter could not be living in themselves, but could only become alive by being permeated with an immortal spirit, the "psyche" (13).

This idea of Plato's played a tremendous role in the later development of our problem. To a certain degree it is reflected in the teachings of Aristotle which subsequently became the chief scientific culture of the Middle Ages and which dominated peoples' minds for almost 2000 years. Accordingly to Aristotle (384–322 B.C.), in addition to the birth of living creatures from like species, spontaneous generation from nonliving matter still occurs and always has occurred (14). Thus, common worms, bee larvae, wasps, as well as ticks, glowworms, and various other insects are born from dew, rotting slime and manure, from dry wood, hair, sweat, and meat. All kinds of worms are born from the decaying parts of the body and excrement. Mosquitoes, flies, moths, May flies, manure beetles, gall midges, fleas, bugs and true lice (partly as such, partly as larvae) emerge from the slime of wells, rivers and seas, from the humus of fields, from molds, manure, from rotting wood and fruit, from animal sewage and dirt of every nature, from vinegar, wastes, and also from old wool (15). However, not only insects and worms, but other living beings also, according to Aristotle, can be spontaneously created. Thus crabs, various mollusks ("testaceans") are born from the moist earth and rotting slime, eels and some other fishes from sea mud, sand, and rotting algae. Even frogs, and under certain circumstances, salamanders, can be formed from coagulated slime. Mice emerge from the moist earth. In a similar manner some higher beings are born, first appearing in the forms of worms. "For this reason, and concerning humans and quadupeds" – wrote Aristotle – "If they at some time were 'earth born' as some people maintain, one may postulate two methods of generation, either from worms which form themselves first, or from eggs." Aristotle not only described the different cases of spontaneous generation in his works; more importantly, he gave to this phenomenon a certain theoretical basis and created his theory of spontaneous generation. In the course of time, apparently, his views changed somewhat, but in the final result they composed the basic idealistic concepts of the origin of life.

Aristotle considered that living beings, as well as other concrete things ("essences") are formed as a consequence of the combination of some passive source "matter" (by these words Aristotle apparently meant what we would not call material) with an active source – "form." The form for living beings was the "entelechy of the body" – the soul. It formed the body and moved it. Thus matter did not have life but was embraced by it, being formed, organized by means of the force of the soul, the purposeful, internal essence of which ("entelechy") brought matter to life and preserved life.

The opinions of Aristotle had a tremendous effect on the whole future history of the problem of the origin of life. By his unquestioned authority, he confirmed the data obtained by direct, naive observation and for many centuries predetermined the future fate of the study of spontaneous generation. All succeeding Greek and Roman philosophical schools shared completely the opinions of Aristotle on the possibility of spontaneous generation of living things. In addition to this, the theoretical foundation of this "phenomenon" acquired a more and more idealistic and even mystical character in the course of time.

Several works of the third and second centuries B.C. contain numerous tales and "miraculous stories" of "plagues of lice," by which the juices of the human body were transformed into parasites; of the origin of worms and insects from rotting materials; of crocodiles from the mud of the Nile, etc. Thus the philosophical school of the Stoics, the most authoritative at this time, considered that animals and plants were generated through the action of engendering force, related to "pneume."

This belief was widely distributed both in the East and in the West by several later Stoic philosophers and writers, Posidonius in particular, who traveled widely in various countries. It received universal recognition in the beginning of our era. In scientific treatises, in political speeches, and in artistic works of this time, we constantly encounter descriptions of various examples of spontaneous generation. We find them also in Cicero, and in the famous geographer Strabo, and in the versatilely educated Philo of Alexandria, and in the historian Diodorus, and in such poets as Virgil and Ovid, and later in Seneca, Pliny, Plutarch, and Apuleius (16).

The concepts of spontaneous generation derived a definite idealistic character from the Neoplatonists (in the third century of our era). The head of this philosophical school, Plotinus, taught that living things could have originated from the earth, not only in the past, but are originating now also in the processes of decay. He explained this phenomenon as a result of the animation of matter — "life-creating spirit" *(vivere facit),* and apparently he was the first to formulate the concept of "life force" which survived even to our day in the teachings of contemporary vitalists (17).

Early Christianity drew guidance from the Bible on the spontaneous generation of life; the Bible in its turn borrowed data from the mystical writings of Egypt and Babylon. The religious authorities at the end of the fourth and beginning of the fifth centuries, the "Christian Church fathers," combined these writings with the teachings of the Neoplatonists and developed on this basis their mystical conception of the origin of life.

Basil the Great, living in the middle of the fourth century, was and still is at the present time one of the chief religious authorities of the Eastern Christian church. It was under his influence that the leaders of Orthodoxy formed their concepts of living nature. His book "Hexaemeron" has been preserved in the church literature, particularly in the Russian language, up to our day. Concerning our problem we can read in it the following: "As one being is produced through the succession of that existing before, another being even now is being created alive from the earth itself. Because not only does it produce grasshoppers in a rainy spell, and a thousand other feathered air-borne breeds, of which many, because of their smallness, have no name, but it also brings forth mice and toads. Around the Egyptian Thebes when during the heat it rains hard, the whole country is overrun with field mice. We see that eels are formed only from slime. They multiply not from an egg and not by other means but derive their origin from the earth" (18).

All these cases of spontaneous generation of living things* occur, according to Basil the Great, through divine command, which has not ceased to operate with unabated strength from the creation of the world to our day.

Just as Basil the Great served as high authority for the Eastern Church, so St. Augustine served the Western Church. He also accepted spontaneous generation of living things as an immutable fact, and in his teachings strove only to base this phenomenon on the world concept as seen by the Christian Church. "Just as God," he wrote, "As a rule creates wine from water and earth through the medium of the grape vine and grape juice, but in some cases, as in Cana in Galilee, He can create it directly from water, so with living things – He can cause them to be born from seeds or create them from non-living matter wherein is lodged the invisible spiritual seed *(occulta semina)*" (19).

Thus Augustine saw in the spontaneous generation of living things the signs of divine judgment – the infusion of life into inert matter by a life-creating soul. He thus affirmed that the teaching of spontaneous generation was completely in accord with the dogma of the Christian Church.

During the whole of the Middle Ages, faith in spontaneous creation dominated the minds of people without exception. At this period

*Many of these cases are apparently borrowed from Aristotle.

philosophical thought could exist only under the mantle of Church doctrine. Every philosophical question could receive universal recognition only when it was connected with a religious question. Philosophy was "a servant of religion" *(ancilla theologiae)*. Questions of natural science were thrust into the background. Nature was visualized not in terms of observation and experiment but in terms of the study of the Bible and religious writings. Only an extremely scanty knowledge of mathematics, astronomy, and medicine penetrated Europe through Arab and Jewish scholars.

In the same way, often by means of very distorted translation, the writings of Aristotle reached European countries. At first his teachings were considered dangerous, but when the Church understood the great usefulness of his teachings for many of their purposes, it elevated Aristotle to the rank of "precursor of Christ in problems of natural science" *(praecursor Christi in rebus naturalibus)*. In this way, according to the apt expression of V. I. Lenin "The scholastics and preachers took the dead from Aristotle and not the living . . ." (19a).

Theologians of the Middle Ages, dealing specifically with the problem of the origin of life, made extensive use of the doctrine of spontaneous generation, the essence of which they viewed as the animation of nonliving matter with an "eternal divine spirit."

Here we can refer to one of the most prominent representatives of scholastic Aristotelianism, the Dominican Albert von Bollstadt, nicknamed the Great (1193–1280). According to legend, Albertus Magnus was an earnest scholar of zoology, botany, alchemy, and mineralogy. However, in his numerous writings he allots considerably less space to independent observation than to material borrowed from ancient authors. On the problem of the origin of life, Albertus Magnus firmly supported the theory of spontaneous generation and in his book *On minerals (De mineralibus)* he especially emphasized the fact that the emergence of living things from putrefaction is a result of the "life-giving force" *(virtus vivificativa)* of the stars.

In his writings on zoology Albertus Magnus gives numerous descriptions of the spontaneous generation of insects, worms, toads, mice, etc., from different types of materials, from moist earth, vapors, sweat, and various kinds of impurities. Precisely the same vapors of earth and water under the influence of heat and the light of the stars breed numerous plants, not only fungi and sponges, but even roots, shrubs, and trees, which often grow in places where seeds cannot have been carried (20).

The same ideas predominate in Albertus Magnus' student Thomas Aquinas (1225–1274) (21). In his chief work "Summa Theologica" considering the problem of the origin of life, he is guided on the one hand, by those views which he ascribes to Aristotle, and on the other, by the teaching of Augustine on the "animating force" *(anima vegetativa)*. He acquiesces completely in the possibility of spontaneous generation of such animals as worms, frogs, and snakes from putrefaction and the action of the sun's heat.

The teaching of Thomas Aquinas to this day is recognized by the Catholic Church as the one true philosophy. Thus the Western Church has supported all through the previous centuries the principle that living things arise from nonliving matter animated by a spiritual source.

The same point of view was also held by the religious authorities of the Eastern Church who in this respect based their opinions mainly on the expressions of Basil the Great. As an illustration may be quoted the Church's opinions concerning the spontaneous generation of animals which were set forth as late as the eighteenth century, by such outstanding figures in the Russian Church as Dimitrii Rostovskii and Feofan Prokopovich. Dimitrii, bishop of Rostov, living at the time of Peter I, in his work "Chronicle Giving Briefly Deeds from Creation to the Birth of Christ" (1708) (22) wrote that Noah did not bring into his ark animals capable of spontaneous generation. They died at the time of the Flood and then arose anew. "Just so those born from earth's moisture, from decay and rot like mice, bullfrogs, scorpions, and next things crawling on the earth, divers worms, bugs and even cockroaches; and mosquitos arising from the heavenly dew and midges and other like things, they all perished in the flood, and then after the flood they were born again from the same substances."

Feofan Prokopovich (23) in the theology course which he taught in the Kiev Theological Academy, develops almost word for word the same idea: "Let us add, that the majority of animals which breed without the copulation of parents, but emerge spontaneously from putrefaction, it was not necessary to shelter in the Ark; such animals as mice, worms, wasps, bees, flies, scorpions."

Even in the nineteenth century Archbishop Veniamin of Nizhegorod in a translation of the book of Francius (24) stated that insects, worms, frogs and mice were spontaneously generated "from putrefying stumps, from animal dung, from sea sand, from the earth's rot, from corpses, etc."

As we have already indicated, natural science in the Europe of the Middle Ages occupied a very low level. It was completely subordinate to religion. The natural phenomena observed by travelers and scholars of this time usually were not only interpreted but even described according to the demands of scholastic wisdom in complete correspondence with ecclesiastical writing and Church dogma. Thus the writings of scholars of the Middle Ages abound in the most fantastic descriptions and sometimes even pictures of the spontaneous generation of various insects, worms, and fish from the slime and moist earth, frogs from the May dew and even lions from the stones of the deserts. Especially characteristic of the methods of perceiving nature prevalent in the Middle Ages were the legends at that time of the goose tree, of the plant lamb, and the homunculus.

We find mention of the goose tree at the beginning of the eleventh century in the writings of Cardinal Peitro Damiani (1007–1072). The English encyclopedist Alexander Neckam (1157–1217) develops the doctrine of the formation of birds from the resin of pine trees on contact with sea salt. Subsequently this doctrine of the plant origin of geese and ducks became so generally accepted that their meat was eaten on fast days, which then had to be prohibited by special dispensation of Pope Innocent III. But in spite of this, even at the end of the fifteenth century (that is, almost three centuries later) Ritter Leo von Rotsmithal describes a dinner given in his honor in London by Herzog Klarenskii, at which ducks were served as a hot dish called "fish" (as a fasting food), the ducks having been spontaneously generated from the sea. However, Rotsmithal notes that these "fish" tasted very much like duck (25).

The famous traveler Odorico di Pordenona (died in 1331) first mentions the vegetable lamb. He told about a "reliable" people, of the Tatar Khanate of Khadli who cultivated tremendous gourds, which were split open on ripening; they contained lambs covered with white wool which had a very tasty meat (26). Mandeville (1300–1372), describing his journey to the eastern countries, also tells of a whole tree, in the melonlike fruit of which live lambs were growing. This tale has been handed down from century to century and even in the middle of the seventeenth century was again repeated by Adam Olearius in his description of his travels to Moscow and Persia. "They told us," he wrote, "that in Samara, between the rivers Volga and Don, there grows a rare form of melon, or rather gourd, which is huge and of a variety resembling the usual melon, in external appearance looking like a lamb, its members having that outline, and this is why the Russians call it 'baranets' (for "sheep"). This vegetable lamb ate the grass around it, and often is taken by wolves, which are great hunters of it." In addition, Olearius wrote, he had himself seen the wool of this "baranet" (27).

ʻThe legend of the homunculus grew up in the wake of alchemical experiments. Apparently it appeared as early as the first century of our era. This legend states that by mixing a passive maternal source and an active paternal source, one could artificially produce the phenomenon of birth and obtain the embryo of a small man – the homunculus.

Like the legend of the goose tree and the vegetable lamb, stories about the homunculus occur throughout all the Middle Ages and are found in a number of alchemical writings. A typical early natural philosopher of the sixteenth century, Theophrastus von Hohenheim, known under the name of Paracelsus (1498–1541), even gives in his writings an "accurate receipt" for the preparation of an homunculus.

Paracelsus in general was a convinced adherent of the theory of spontaneous generation of living beings. He believed that an active life force – Archaei – predominates in the body of animals and man; this force can be controlled by means of certain magic rules. It determines the formation of the organism and its future behavior. In accordance with these philosophical views, Paracelsus evolved an hypothesis on the spontaneous generation of life. He even carried out a series of his own observations on the sudden generation of mice, frogs, toads, and turtles from water, air, straw, rotting wood, and every form of waste product (28).

In the second half of the sixteenth century and especially in the seventeenth century, observations on the phenomena of nature became more accurate. Copernicus (1473–1543), Bruno (1548–1600), Galileo (1564–1642) demolished the old Ptolemaic system and formed a correct representation of the world of stars and planets surrounding us (29). But this flowering of accurate knowledge still did not apply to biological problems. The idea of the primordial spontaneous generation of living things remained in full strength in the minds of the investigators of the day.

The well-known Brussels doctor Van Helmont (1577–1644) can here be cited as an example. He was so completely a master of the methods of accurate experiment that he could closely approach a solution to the complicated problem of plant nutrition, but in spite of this, he considered completely indisputable the theory of spontaneous generation of living things. Moreover, he based his position on a series of observations and experiments. Van Helmont himself is the author of a widely known recipe for obtaining mice from grains of wheat. Since human vapors, in his view, were a breeding source, it was only necessary to put a dirty shirt in some sort of vessel containing wheat grains. After 21 days the "fermentation" is broken off and the emanations of the shirt, together with the emanations of the grain form living mice. Van Helmont was very much surprised that these artificial mice which he obtained were entirely the same as natural mice produced from paternal sperm (30). The creator of the theory of blood circulation, Harvey (1578–1657), also did not deny arbitrary spontaneous generation. Although he is the author of the famous phrase, "Everything living comes from the egg" *(Omne vivum ex ovo)*, he attributed to the word "egg" here a very broad interpretation and considered entirely possible the arbitrary spontaneous generation *(generatio aequivoca)* of worms, insects, etc. as a result of the action of special forces developing during putrefaction and analogous processes (31).

This same point of view was held by a contemporary of Harvey's, the founder of seventeenth century English materialism, Francis Bacon (1561–1626). In his writings he expressed the opinion that different plants and animals, (flies, ants and frogs, for example) could emerge spontaneously during the rotting of various materials. But he approached this phenomenon from the materialistic viewpoint and saw in it only an indication of the absence of an impenetrable barrier between the inorganic and organic worlds (32).

An especially clear expression of the materialistic interpretation of spontaneous generation is found in the teachings of Descartes (1596 –1650) (33).

This great French philosopher, although he did not doubt the spontaneous generation of living things, categorically rejected the position that this generation occurs under the influence of the "anima vegetative" of the scholasticists, the "Archaei" of Paracelsus, or the "spirit of life" of Van Helmont or any sort of spiritual source. In sharp contrast with the religious scholars and with the anthropocentric tendencies of medieval natural philosophy, the physicist Descartes attempted to turn all the qualitative multitude of natural phenomena into a question of matter and its movement.

From Descartes' point of view, the living organism does not need to be explained by any special obedience to a "vital force"; it is nothing

other than a machine, extremely complex in its structure, but completely understandable; its movement depends exclusively on pressure and the
impact of particles on it similar to the movement of a wheel in a tower
clock. This is why various living things can be spontaneously generated
from the nonliving matter surrounding them. Specifically, every possible
plant and animal such as worms, flies, and other insects emerge from the
exposure of moist earth to the sun's light or from the phenomena of
putrefaction. But the intervention of some sort of "spiritual source" is
not required. Spontaneous generation is only the natural process of
autoformation of those complex machines—living things—a process which
always occurs under predetermined conditions, although it is true, they
are as yet only partly understood.

Thus by the middle of the seventeenth century the theory of spontaneous generation of living things had not been doubted by anyone.
The dispute between the mystical doctrines of the Middle Ages and
briskly developing materialism applied only to the theoretical interpretation of this phenomenon: whether to regard spontaneous generation as
the manifestation of a "spiritual genesis" or as the natural process of
autoformation of living things.

However the more and more extensive and accurate investigations of
living nature began to shake peoples' faith in the authenticity of the
"fact" of spontaneous generation.

The turning point in this respect may fairly be considered the
experiments of the Tuscan doctor Francesco Redi (1626–1698) (34)
(Fig. 2). Redi had the honor of being the first to come forward with
experimental refutation of spontaneous generation so firmly established
for many centuries. In his paper "Esperienze intorno alla generazione
degl'inetti" (1668) this scholar describes a series of experiments he
carried out which indicate that white maggots in meat are non other
than the larvae of flies. He placed meat or fish in a large vessel covered
with a very fine Neapolitan muslin, and for yet more careful protection
covered the vessel with a muslin cloth on a frame. Although many flies
lit on the cloth, worms did not appear in the meat itself. Redi stated
that he had observed that the flies laid their eggs on the cloth and that
only when the eggs fell through onto the meat did they develop into
maggots. He concluded from this that the decaying substance was only a
place or nest for the development of the insects, and that the necessary
prerequisite for their appearance was the laying of eggs, without which
the flies never would emerge.

FIG. 2a. Medal with a portrait of Francesco Redi.

But one must not think that even Redi succeeded completely in ridding himself of the notion of spontaneous generation. In spite of his brilliant experiments and his correct interpretation of them, he wholly accepted the possibility of spontaneous generation in other cases; thus, for example, he believed that intestinal and wood worms emerged spontaneously from rotting materials; and even more that worms which are found in oak galls are formed from plant juices.*

This example makes it clear that convictions accumulated over the centuries (even wrong ones) are not easily changed. During the whole of the eighteenth century and even into the beginning of the nineteenth, many scholars and philosophers of different persuasions and schools and even writers and poets often depicted various fantastic examples of spontaneous generation in wild beasts, fish, insects, and worms, or documented fully, from their point of view, the possibility of such phenomena. Only very gradually, as a consequence of more accurate observations of living nature and especially more detailed knowledge of

*It was only later that this concept was refuted by the investigations of the Paduan medic and naturalist Vallisnieri (1661–1730).

FIG. 2b. Lazzaro Spallanzani. (See page 25).

the structure of living things, was the impossibility recognized of the generation of such complicated forms from unstructured mud and rotting matter. Faith in spontaneous breeding of all highly organized living things was thus expelled from scientific custom. But the idea itself of primary genesis did not disappear and on the contrary was further developed in the eighteenth and nineteenth centuries in relation to the simplest and tiniest of living things, the microorganisms.

After the Dutch investigator Anthony van Leeuwenhoek (1632–1723) (35) discovered a new world of tiny living creatures invisible to the naked eye with a magnifying glass which he made himself, these animalcules were found wherever putrefaction or fermentation of organic substances had occurred. Microbes were found in different types of plant extracts and decoctions, in decaying meat, in spoiled broth, in sour milk, in fermenting wort, etc. If a quickly perishable or easily spoiled substance were left for a time in a warm

FIG. 2c. Louis Pasteur. (See page 26).

place, microscopic living things which had not been there earlier soon began to develop. With the faith in arbitrary spontaneous generation prevalent at that time it reasonably followed that just here, in the decomposing broth and decoctions, the spontaneous generation of living microbes from nonliving matter occurred. This idea was very authoritatively maintained by the noted German philosopher Leibniz (1646–1716) (36) and the French naturalist Buffon (1707–1788) (37).

The Scottish clergyman and naturalist Needham (1713–1781) (38) even carried out special experiments for the purpose of demonstrating the phenomenon of "spontaneous generation of microbes."

"I took," wrote Needham, "a quantity of mutton gravy hot from the fire and shut it up in a phial closed with a cork so well matched that my precautions amounted to as much as if I had sealed my phial hermetically." After this Needham also carefully heated the vessel on hot ashes. But in spite of all this, after several days the vessel swarmed with microorganisms. He made similar observations with various organic liquids and broths. He naturally concluded that it was entirely possible and even obligatory to obtain spontaneous generation of microorganisms from rotting organic substances.

However, these experiments of Needham's were subjected to cruel criticism on the part of the Italian scholar Spallanzani (39). Spallanzani, just as did Needham, carried out experiments with the object of establishing or refuting the possibility of spontaneous generation. He came to a directly opposite conclusion, however, on the basis of his experiments. He asserted that Needham's experiments succeeded in producing microorganisms only because of insufficient heating of the vessels containing the liquid and that consequently they were incompletely sterilized. He performed hundreds of experiments in which plant decoctions and other organic liquids were subjected to greater or lesser periods of boiling, after which the vessel containing them was sealed, thus impeding the access of air to the liquid; the air, according to Spallanzani's hypothesis, being the bearer of the microbial embryos. In all these cases when the operations were performed with proper care, the liquid placed in the vessel did not spoil and no living things appeared in them. Nevertheless Spallanzani did not succeed in convincing his contemporaries that he was right.

The doctrine of self-induced generation was later defended by many naturalists and philosophers at the end of the eighteenth and beginning of the nineteenth century, particularly by the German idealistic natural philosophers. Although Immanuel Kant himself (1724–1804) (40) considered that the primary intrinsic cause of the genesis of organisms lay in the suprasensible (metaphysical) and thus the hypothesis of spontaneous generation was only a "bold adventure of the intellect," the later natural philosophers Hegel (1770–1831) (41), Schelling (1775–1854) (42), and Oken (1779–1851) (43) in their writings extensively developed the theme of spontaneous generation of life *(generatio equivoca)*. In addition to these philosophical expressions, in the first half of the nineteenth century, a series of experiments were performed with the object of proving or disproving the spontaneous generation of microbes; these include the experiments of Schwann (44), Schulze (45), and Schröder and Dusch (46). But in spite of the fact that they all in general refuted the theory of spontaneous generation, their proof was insufficient. In some cases, for obscure reasons, microorganisms appeared in the liquids. Now we know that this was a result of some accidental technical error, but contemporaries assumed otherwise. According to the desires of believers in spontaneous generation, the appearance of microbes easily could be interpreted and actually was interpreted as indicating that spontaneous generation can occur; not always, but under certain circumstances. This idea was adhered to even by such outstanding investigators as Dyuma, Naegeli and other scientists of the midnineteenth century.

The struggle of ideas on the question of spontaneous generation of microorganisms attained its highest pitch when in 1859, Pouchet (47) published a paper in which he tried to prove the theory of spontaneous generation. About one hundred years separate us from the experiments of Pouchet, but when we examine these experiments we are amazed at how crudely they were set up and what degree of contamination they permitted. However Pouchet's work made a great impression on his contemporaries. The French Academy of Science established a prize to be given to the man who could shed light on the problem of the primary generation of living things with precise and conclusive experiments. The prize was awarded to Louis Pasteur (1822–1895) (48), who in 1862 published his work on arbitrary spontaneous generation. In a series of brilliant experiments, absolutely irrefutable, he proved that the formation of microorganisms from various broths and solutions of organic substances was impossible. Pasteur accomplished this only because he departed from the well-trodden path of blind empiricism and broadly encompassed the whole problem in his work. He gave a rational explanation for all the earlier experiments and indicated the sources of error in his predecessors' work. He refuted the earlier hypothesis in which putrid broths and decoctions bred microbes, and showed that, on the contrary, the putrefaction and fermentation in themselves of these liquids were a result of the life activity of the microorganisms, the embryos of which have been brought from outside. All attempts to disprove this statement and to find an example of spontaneous generation of microbes were in vain. From our modern point of view, this is entirely understandable since we know that microorganisms are not simple aggregates of organic substances as was thought before Pasteur's time. A detailed study of these primitive living things has shown that they are extremely complex and complete organizations of matter. It is absolutely impossible to assume now that such a structure can emerge directly before our eyes from nonstructured solutions of organic substances. This is essentially as absurd as the assumption that frogs are bred from dews of May or lions from the stones of the desert.

Pasteur's investigations made a tremendous impression on his contemporaries. And of course this should be expected. Pasteur's scientific feat in essence can be compared with that of Copernicus. Here also prejudices were disproved which had dominated the minds of people for a thousand years. However, these investigations did not serve as a means for solving the problem of the origin of life. On the contrary, they brought about a serious crisis in natural science concerning this problem not only at the end of the century but also into the entire first half of our century. The reason for this crisis lay not in the existence of the problem itself (as the vitalists attempted to show) (49), but in the methodologically fallacious mechanistic means of understanding the existence of life itself, which dominated natural science at that time.

Mechanism as a philosophy never has recognized and even now does not recognize any important qualitative difference between the inorganic world and the world of living things. A direct, sudden change from one world to another—the spontaneous generation of living things—is entirely necessary to this point of view. The simplest organisms must arise from nonliving matter in a manner similar to that in which a crystal having a certain internal structure is formed from the nonstructured mother liquor. Thus, primary spontaneous generation of organisms is logically required as a consequence of the mechanistic view of life (50). This explains these persistent and very passionate attempts to discover the phenomenon of spontaneous generation which have been made over a long period of time, even after Pasteur's experiments (51).

However, all these attempts came up against the completely insurmountable wall of the facts (52). All communications that have reported examples of spontaneous generation have proved false. They were the result of incorrect experimental procedures or the wrong interpretation of data and were refuted after further checking. This knocked the ground out from under the feet of those investigators who saw in spontaneous generation the one and only way to solve the problem of the origin of life. Devoid of the opportunity to experiment along those lines, they were deeply disappointed and in fact have made no attempts to resolve this "accursed" problem and only have sought some reason for its unresolvability.

Two points of view had formed by the beginning of our century. One doctrine, essentially maintaining the previous position of spontaneous generation, considered it to be not a regularly occurring phenomenon but an extremely rare "lucky accident" which could occur perhaps only once during the whole existence of the Earth and hence was unverifiable by experiment. The other doctrine, on the contrary, completely denied the possibility not only of spontaneous generation but of the genesis of life in general; it sought to explain the emergence of our earthly life by the transport of seeds from other worlds which "may be regarded as eternal repositories of living forms, as perpetual plantations of organic germs" (53).

"It appears to me," wrote Helmholtz in this connection (54), "to be a fully correct procedure, if all our efforts fail to cause the production of organisms from nonliving matter, to raise the question whether life has ever arisen, whether it is not just as old as matter."

Engels (55) at the end of the last century gave a correct and crushing criticism of similar opinions. He stated that two completely different concepts were being confused: (1) the perpetuity of life as something never being generated, but only successively being transformed from one organism to another, and (2) the constant "perpetual" genesis of life as a special form of the movement of matter, being always and everywhere renewed as long as the proper conditions exist.

We now are confident that our planet is not the only one inhabited by life, that this form of movement of matter has arisen and will arise on some other suitable objects in the universe. But the transport of the germ of life to Earth from other worlds, even if it were in fact found to exist (which is very unlikely) would not furnish a solution to the problem of the genesis of life, since the genesis must have taken place on the Earth equally, as well as on other corresponding celestial bodies.

Nevertheless, the search for such a transport occupied the minds of scholars at the end of the last and the beginning of our century. Most often, the hypothesis has been advanced that the germs of life were brought to Earth by meteorites or combined with cosmic dust (the hypothesis of panspermia). The first hypothesis was developed by Richter (1865) (56) and then supported by Thomson (57), Helmholtz (58), Van Tieghem (59), and others. According to this hypothesis, meteorites traveling through the Earth's atmosphere are strongly incandescent only on the surface, while the interior remains cold. Thus the embryos of organisms inhabiting meteorites or the planets from which they were formed were preserved alive in the interior of the meteorites.

However, numerous attempts to determine directly viable embryos or at the very least their dead residues in meteorites have not given positive results. According to the evidence of Meunierref (60), Pasteur tried to isolate viable bacteria from a carbonaceous meteorite but obtained negative results and therefore did not publish them. Considerably later (1932), Lipman (61) stated that he had isolated from the sterilized surface of meteorites microbes which were capable of growing in nutritive media. However, these microbes were entirely similar to terrestrial microbes, and apparently were picked up as the meteorite fell to the Earth. As recent model experiments of Imshenetskii (62) have shown, this occurs easily due to the formation of deep fissures in the meteorites (carbonaceous chondrites) which sometimes reach clear to the center. Water, accompanied by terrestrial contaminants (not only bacteria but even pollen of higher plants) is sucked into the fissures as the meteorite falls to the Earth.

Even in our time (1962), Nagy, Claus, and Hennessy (63) have published a communication stating that structured formations can be found in the Orgueil and Iruna carbonaceous chondrites which are residues of organisms formerly inhabiting the meteorite material, according to the opinions of these authors. However, on more careful checking, these opinions have not been confirmed. As Briggs (64) and also Anders (65) have shown, the structured formations of Nagy and others are mineral granules which only imitate the biological formations.

The second hypothesis of panspermia was developed in detail in the beginning of our century by the famous Swedish physical chemist Arrhenius (66). This scholar considered that spores of life together with particles of cosmic dust could be transferred from one heavenly body to another under the pressure of stellar rays. Arrhenius calculated the rate with which this transfer would have to be accomplished and presented a series of reasons proving, in his opinion, the possibility of the transport of the spores of life through cosmic space in a viable condition. However, this latter statement met with very serious objections on the part of later investigators (67,68). An especially serious danger for living embryos is short wavelength ultraviolet radiation, penetrating interplanetary and interstellar space and destructive of all life. Modern cosmic flights can solve directly the question of the presence of life beyond the limits of the terrestrial atmosphere. But until that occasion we have no positive data.

The penetration of man into the cosmos has created the daring hypothesis that the initial forms of life were carried to our planet at some time by cosmonaut or astronaut visitors, highly conscious beings, interplanetary and interstellar travellers. However, this hypothesis, widely current mainly in popular scientific and fantastic literature, has as yet no factual basis. It is interesting to note that the outstanding Russian scientist and inventor Tsiolkovskii (69), in spite of his ardent belief in the possibility of interplanetary voyages, categorically denied such artificial transport of microbes. In his posthumous manuscript (1919) "Origin of Plants on the Earth and their Development," we read the following: "My work showed that artificially, by special methods which mankind will not soon master, a living being can be sent without danger from Earth to another planet and return." But such a transfer of life "with the assistance of intelligence" had not taken place in the past in Tsiolkovskii's opinion for no traces had been observed suggesting that at any time or place there have been such highly developed beings deliberately visiting the Earth. "You know," concludes Tsiolkovskii, "Life has not migrated to Earth from the planets even by means of intelligence." However, now when the prospects of flight to the Moon and to the planets nearest to us have already become real, it is necessary to prevent the contamination of these celestial bodies with terrestrial organisms, since this contamination would strongly interfere with solving the problem of the genesis of life beyond the limits of the Earth.

Proponents of spontaneous generation, in addition to assertions that such a genesis of primordial organisms is the rarest "lucky accident," also developed the idea that this genesis required some set of special conditions which existed some time in the past and now are lost to Earth.

This doctrine was formulated most distinctly in the second half of the last century by the leading German naturalist of his time, Haeckel (70). Haeckel considered that the most primitive organisms would have to be generated spontaneously at some time from inorganic matter as a result of the formative action of some sort of special external physical force. This, according to Haeckel, does not contradict the circumstance that in the present time we do not observe the spontaneous generation of microbes. This is because of the fact that these forces are now absent in the nature surrounding us, but at some time they occurred on the surface of our planet.

Haeckel himself did not consider it possible to represent these forces in detail with the knowledge existing at his time, but later scholars, supporting this point of view, enlisted in their reasoning such forces as

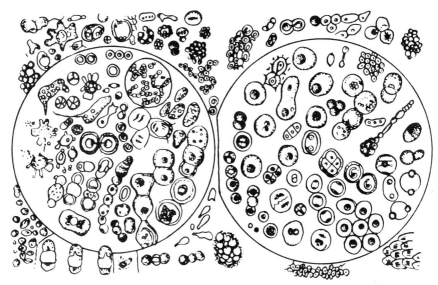

FIG. 3. The artificial "cells" of A. Herrera.

electrical charges, ultraviolet light, and special forces of a chemical nature, then later the corpuscular radiation of radioactive elements. As we will see below, all these factors actually must have played an important role as the sources of energy for transforming organic substances in the process of their evolution on the primitive Earth. But alone, as such, they could not cause spontaneous generation of organisms either in the remote past or in the present.

Thus all attempts to reproduce spontaneous generation of organisms from nonliving materials by means of these forces in a laboratory setting naturally concluded in complete failure. All such experiments, including the creation of the "viable granules" of Dubois (71), the "cells" produced by radium of Kuckuck (72), the "plasmogens" of Herrera (Fig. 3) (73), the synthetic "fungi and algae" of Leduc (74) are extremely primitive and naive in character. Knowledge of them to a certain extent leads us to understand why an entirely negative attitude was formed in the biological literature of the second and third decades of our century, toward the problem of the origin of life; its solution was not considered worth the time of a serious investigator.

It seemed that the metaphysical approach to the problem of the origin of life had led natural science to a hopeless blind alley. Quite different prospects opened up before scholars as a result of using the evolutionary method.

The significance of this method as applied to our problem apparently was evaluated even as early as Lamarck (75). In his work "Analytical System of Positive Knowledge of Man" published in 1820, Lamarck regarded the genesis of life from the nonliving as a process of the gradual development of matter.

Engels in the 1870's indicated very positively that the only possible route for the origin of life was the evolutionary development of matter (55). Life, according to Engels, does not exist eternally and does not originate spontaneously and suddenly, it emerges in the process of the evolution of matter everywhere and always, when the necessary conditions arise for this sort of evolution.

However, the great significance of Engels' position did not find a sufficiently wide reflection in the works of naturalists of his time. Only a very few of them, and then only in general terms, recognized the significance of the evolutionary principle for the solution of the problem of the origin of life. The address of the distinguished Russian botanist and cytologist Belyaev (76) at Warsaw University in 1893 should be noted here, as well as the speech given by Schäfer (77) at the annual meeting of the British Association in Dundee in 1912. His concern in this speech was the origin of life. Schäfer said, "We are not only justified in believing, but are compelled to believe, that living matter must have owed its origin to causes similar to those which have been instrumental in producing all other forms of matter in the universe, in other words to a process of gradual evolution. . . . Looking, therefore, at the evolution of living matter by the light which is shed upon it by the study of the evolution of matter in general, we are led to regard it as having been produced not by a sudden alteration whether exerted by a natural or supernatural agency, but by a gradual process of change from material which was lifeless, through material on the borderland between the animate and inanimate, to material which has all the characteristics to which we attach the term 'life'."

Timiryazev (78) thought very highly of these statements of Schäfer's and subscribed to his opinions completely. But neither Schäfer himself or any other of the scientists of that time could develop any sort of concrete hypothesis on the means of the evolution which they advocated. On the contrary, Schäfer considered this sort of hypothesis still unworkable at the level of knowledge available to him.

The main obstacle which immediately stood in the way of any attempt to establish firmly the begetting of life by evolution was the conviction generally held at that time that under natural conditions organic matter *could* be formed and essentially *was* formed only by biological means, only by organisms.

It is true that at the end of the last and beginning of the present century the experiments of Wöhler (79), Butlerov (80), Berthelot (81), and a whole pleiad of other chemists on the synthesis of various organic substances were well known. But this did not in any way change the opinions stated above. Let us consider that in this case these syntheses are also accomplished by a living being – man, who consciously carries out a specific order of chemical reactions which do not exist under natural conditions outside living beings.

In addition to this, all direct observations under natural conditions have always accordingly and simply revealed only the biogenic pathway of organic matter on the Earth's surface. The whole contemporary living population of our planet now exists only at the expense of those organic substances which are the base of the material substrate of all living creatures, formed in the process of photosynthesis or chemosynthesis of green plants or specific bacteria. Moreover, even the mineral organic substances of oil and coal are basically biogenic in their origin, since they are chiefly products of the extensive decomposition and changes in residues of organisms existing at some time and heated in the Earth's core. The conclusion may be drawn from this that formation of organic substances under natural conditions is possible in general only through the medium of organisms (biogenically). This assertion naturally gave rise to an insurmountable difficulty in solving the problem of the origin of life. Organic substances consist basically of the material substrate of all living creatures. There is no life on our planet without them since only regular interaction of these substances can bring about the certain dynamic organization of living bodies which is the essence of biological metabolism. Thus it is understandable that a study of the evolution of matter as a means to determine the origin of life must begin with the solution of the problem of how the most primitive representatives of organic compounds were formed in the beginning. But if these substances can be formed under natural conditions only in the process of the life activity of organisms, we unwillingly fall into a vicious circle from which there is no exit.

In my report delivered in 1922 to the Russian Botanical Society (82),

I strongly stated that such a "no exit" position is brought about only if we limit our investigations to the study of the situation existing on the Earth in the modern epoch. If we widen the circle of our investigations beyond the limits of our planet and learn the facts pertaining to other heavenly bodies, we must reach other conclusions.

This type of investigation shows that the simplest organic substances can be found on many diverse objects of our celestial world. These organic substances undoubtedly were formed outside of any sort of connection with life (abiogenically). This permits us to theorize that our planet also is not absolutely excluded in this respect and that the abiogenic formation of organic substances occurred in that period of time when it was still lifeless; only later, with the emergence of life, was this process complicated more rapidly and widely by the perfection of biogenic syntheses.

In 1924 my little book "Origin of Life" was published (83).* I first stated in it, still very schematically, of course, that view which in a more developed form has been my theme in subsequent publications (84). In particular I sought to show the possibility of initial abiogenic formation on our planet of the simplest organic substances – hydrocarbons. The evolution of these substances must have led to the formation of proteinlike compounds, and then to colloidal systems capable of stepwise improvement in their internal structure as the result of the action of natural selection.

Somewhat later (1929), a paper by Haldane (85) was published which had great significance in the development of knowledge on the origin of life. The author also demonstrated that the genesis of organic materials preceded the formation of primitive organisms; he developed an evolutionary outlook on this process.

The theory of the primary formation of organic substances on Earth became very widely disseminated in scientific circles, in particular after the presence of hydrocarbons was discovered in the atmosphere of the large planets and on other cosmic objects. This was greatly aided by the work of several astronomers, physicists, chemists, and geologists. In this connection, one must note Urey's book "Planets, their Origin and Development" (86) and Bernal's "Physical Basis of Life" (87).

Solution of the problem of the further evolution of the primary hydrocarbon compounds on our planet, their transformations under the

*In J. Bernal's book *The Origin of Life* published in 1967 a complete English translation of this book is given as an appendix.

conditions of the ancient hydrosphere and atmosphere of Earth were greatly aided by numerous experiments reproducing these conditions in a laboratory environment. These experiments were successfully begun by the American scientist Miller (88) who obtained amino acids — those most important components of the protein molecule — by subjecting a gas mixture consisting of methane, ammonia, hydrogen, and water vapor to electric discharges. Analogous syntheses of amino acids were carried out by Pavlovskaya and Pasynskii (89) by the action of ultraviolet light.

Subsequently, similar investigations were carried out by numerous scientists in several countries of the world. We will describe these investigations in further accounts. Here it is only necessary to note that a large number of differing substances, as well as their polymers, analogous to proteins, nucleic acids, etc. were synthesized from primitive hydrocarbon compounds. In addition to this, a broad investigation was made of the primary formation from these polymers of multi-molecular formations, the emergence of primitive metabolism in them, etc.

In the middle of our century a decisive crisis in the relationship of naturalists to the problem of the origin of life has occurred.

Earlier this problem was, if it can be thus expressed, prohibited; the scientific literature devoted to it at that time was extremely scanty and in most cases was only general and speculative in character. In contrast to this, the number of investigations in this region during the last 10–15 years has increased rapidly. As a result of this it has even been necessary to bring together investigators in different specialties and to inter-correlate their work. This sort of unification was accomplished at the Moscow Symposium on the Problem of the Origin of Life held in 1957 by the Academy of Sciences of the USSR on the initiative of the International Biochemical Union (90).

This symposium played an important role in our further progress toward a solution of the problem of the origin of life. It not only provided a summing up of the previous investigations; it also pointed out ways for coordinating the work of scientists of various countries and disciplines.

A worthy successor to the Moscow Symposium was the conference on the theme "Origins of Prebiological Systems" (91), which was organized by S. Fox at Wakulla Springs, Florida in 1963. This conference showed that the problem of the origin of life has already passed beyond the stage of general speculative discussions. Now its solution will be accomplished on the basis of strictly scientific observations and experiments. It is no longer a question of hopeless attempts to find sudden spontaneous generation of organisms (as was true earlier) but of study and experimental reproduction of completely feasible phenomena regulated by orderly progression and emerging in the evolutionary development of matter.

It is true that we do not now see in nature surrounding us the processes of the primary origin of life, since the evolution of carbonaceous formations is irreversible and unidirectional in character. We also cannot reproduce artificially the whole of this process in the form in which it occurs in nature since the process lasted for more than a billion years. However, separate steps in the process are completely suitable for objective scientific investigation. At the present time the following three methods are successfully used:

1. Investigation and study of the separate links of evolutionary process under natural conditions when these conditions are the same as they were in the prebiological epoch (that is, before the genesis of life). This method of investigation is extended not only to our planet but also to other objects in the universe, a study of which takes on more and more significance the farther man penetrates the cosmos.

2. Artificial reproduction in the laboratory environment of those conditions which existed on a still lifeless Earth, at separate stages in the evolution of carbon compounds; study of the processes contiguous with these conditions — synthesis of more and more complicated organic substances and the formation from them of polymolecular systems, and, finally, construction and study of models of these systems and processes.

3. Comparative biochemical investigation of modern organisms in different stages of development. These investigations permit us to clarify the order of evolution of metabolism and its earliest links evolved in the very process of the origin of life. Data obtained in this way are the starting points for the construction of artificial models having a more and more complex metabolism.

Successes already presently attained by the use of these methods permit us to hope that the time is not far off when we will succeed in artificially reproducing the simplest forms of life, modeling in shortened form the whole generative process, substituting for long natural evolution the conscious creation of systems and the direct combination of processes which are somewhere successive links in the evolutionary chain. It is more than ever necessary, then, if such an artificial synthesis of life to to be successful to obtain accurately proven and experimentally verified data which will reveal to us how the natural origin of life was accomplished on the Earth's surface.

REFERENCES

1. E. Lippman. "Urzeugung und Lebenskraft." Springer, Berlin, 1933; Arbman, *Arch. Religionswissenschaft* **29**, 293-320 (1931).
2. Bodenheimer. "Materialien zur Geschichte der Entomologie," Vol. 1, p. 24. Berlin, 1928.
3. Denssen. "Geschichte der Philosophie." Leipzig, 1908.
4. I. Growther. "Science Unfolds the Future." Muller, Ltd., London, 1955.
5. Brugsch. "Religion und Mythologie der alten Agypter." Leipzig, 1891.
6. W. Shakespeare. "Antony and Cleopatra,"Act II, Scene 7.
7. "Istoriya filosofii" ("History of philosophy"). Vol. 1. Izd. Akad. Nauk SSSR, Moscow, 1957.
8. A. Makovel'skii. "Doskratovskaya filosofiya" ("Pre-Socratic Philosophy"), Vol.1, No.1. University Press, Kazan', 1918.
9. P. Tannery. "Pour l'histoire de la science helline. De Thales à Empedocles Paris" (2nd ed. by A. Díes). 1930; T. Gompertz. "Griechische Denker. Geschichte der antiken Philosophie," 3rd ed., Vol. 1, Leipzig, 1911.
10. A. Deborin. "Kniga dlya chteniya po istorii filosofii" ("Readings in the History of Philosophy"), Vol. 1. Izd. "Novaya Moskva," 1924-1925, Democritus. Fragments.
11. A. Makovel'skogo. See "Materialisty Drevnei Gretsii" ("Materialists of Ancient Greece"). Gospolitizdat, Moscow, 1955. "Letters and Fragments from Epicurus" (Translated from the Greek). Epicurus: Letter to Herodotus.
12. Lucretius Carus. "On the Nature of Things" ("De rerum natura").
13. Rodemer. Lehre von der Urzengung bei den Griechen und Römern. Dissertation, Univ. Press, London and New York, Giessen, 1928.
14. Simplicius on Aristotle's De Caelo Ancilla to the Pre-Socratic philosophers. Oxford, 1952; Aristotle. "On the Genesis of Animals" (Translated from the Greek).
15. E. Zeller. "Die Philosophie der Griechen." Reisland, Leipzig, 1923.
16. Apuleius. "Metamorphoses."
17. H. Meyer. "Geschichte der Lehre von den Keimkraften." Haustein, Bonn, 1914.
18. St. Basil. "Letters and Select Works" (Transl. by Blomfield Jackson), p.102. Oxford Univ. Press, London and New York, 1895.
19. Cited by E. Lippman. "Urzeugung und Lebenskraft." Springer, Berlin, 1933.
19a. V.I. Lenin. "Philosophical Notebooks," p.332. I.M.E.L., Moscow, 1938.
20. I Sighart. "Albertus Magnus, sein Leben und seine Wissenschaft." Regensburg, Manz, 1857.
21. A. Shtekl'. "Istoriya srednevekovoi filosofii" ("History of Medeival Philosophy"). Izd. Sablina, Moscow, 1912.
22. D. Rostovskii (D. Tuntalo). "Letopisy, skazuyushchaya vkrattse deyaniya ot nachala mirobytiya do rozhdeniya Kristova, sobrannaya iz bozhestvennogo

38   GENESIS AND EVOLUTIONARY DEVELOPMENT OF LIFE

pisaniya i iz razlichnykh khronografov i istoriografov grecheskikh, slavyanskikh, rimskikh, pol'skikh is inykh" ("Chronicle, Giving Briefly Deeds from Creation to the Birth of Christ, Collected from Religious Writings and their Various Greek, Slavic, Roman, Polish and Other Chronographs and Historiographs"). Works, Vol. 4. Moscow, 1857.

23. Th. Prokopowicz. "Christianae orthodoxae theologiae in Academia Kiowiensi," Vol. 1. Lipsiae, 1782; see also S. Sobol'. "Istoriya mikroskopa i mikroskopicheskikh issledovanii v Rossii v XVIII veke" ("History of the Microscope and Microscopic Investigations in Russia in the 18th Century"). Izd. Akad Nauk SSSR, Moscow-Leningrad, 1949.
24. V. Francius. See B. Raikov. *Zh. Min. Narod. Prosveshch.* No. 11, Sect. 3 (1916).
25. Cited by E. Lippman. "Urzengung und Lebenskraft." Springer, Berlin, 1933.
26. A. Tschirch. "Handbuch der Pharmakognosie." Tachnitz, Leipzig, 1909.
27. A. Olearius. Podrobnoe opisanie puteshestviya Golshtinskogo posol'stva v Moskoviyu i Persiyu v 1633, 1636 i 1639 godakh (Detailed description of the journey of the Holstein ambassador into Muscovy and Persia in 1633, 1636 and 1639) (Translated from German); see A. Oleorius. In "Die erste Deutsche Expedition nach Persien (1635-9)" (H. von Staden, ed.). Leipzig, 1927.
28. E. Darmstaedter. "Acta Paracelsica." Munchen, 1931.
29. G. Gurev. "Sistemy mira" ("Systems of the World"). Izd. Akad. Nauk SSSR, Moscow-Leningrad, 1940.
30. W. Bulloch. "History of Bacteriology." Oxford Univ. Press, London and New York, 1938.
31. T.Meier-Steineg and K. Sudhoff. "Geschichte der Medizin," 2nd ed. Jena, 1922.
32. F. Bacon. "The Works of Francis Bacon," Vol. 1. Reever, London, 1879.
33. R. Descartes. "Oeuvres philosophiques." Aime et Martin, Paris, 1838.
34. F. Redi. "Esperienze intorno alla generazione degl'insetti." 1668.
35. A. van Leeuwenhoek. "Arcan naturea detecta Delphis Batavorum." 1695.
36. G. Leibnits (G. Leibniz). *In* "La Monadologie" ("Opera philosophica") (Y. Erdmann, ed.). Berlin, 1840.
37. E. Nordenskiold. "Die Geschichte der Biologie." Fischer, Jena, 1926.
38. T. Needham. *Phil. Trans. Roy. Soc. London* No. 490, 615 (1749); "Idée sommaire ou vue générale du système physique et metaphysique." Bruxelles, 1776; W. Bulloch. "History of Bacteriology," p. 74. Oxford Univ. Press, London and New York, 1938.
39. L. Spallanzani. "Saggio di osservazione microscopiche concernenti il systema della generazione dei sig. di Needham e Buffon, Modena, 1765, Opuscolidi fisica animalee Vegetabile." Modena, 1776.
40. I. Kant. "Kritik der Urtheils Kraft, Lagade und Friedrich." Berlin-Liban, 1790.
41. G. Hegel. "Encyclopedia of the Philosophical Sciences" ("Enzyklopädie der philosophischen Wissenschaften"). Heidelberg, 1817.
42. F. Schelling. "Zeitschrift fur spekulative physik." Gabler, Jena, 1800.
43. L. Oken. "Lehrbuch der Naturphilosophie." Schulthess, Zurich, 1843.
44. T. Schwann. *Ann. Phys. Chem.* 41, 184 (1837).
45. F. Schulze. *Ann. Phys. Chem.* 39, 487 (1836).
46. H. Schröder and T. Dusch. *Ann. Chem.* 89, 232 (1854).

47. F. Pouchet. *Compt. Rend.* **47**, 979 (1858); **48**, 148 and 546 (1859); **57**, 756 (1863); "Heterogenie ou traité de la génération spontanée basée sur de nouvelles expériences." Bailliere et Fils, Paris, 1859.

48. L. Pasteur. *Compt. Rend.* **50**, 303, 675 and 849 (1860); **51**, 348 (1860); **56**, 734 (1863); *Ann. Sci. Nat.* **16**, 5 (1861); *Ann. Chim. Phys.* [3] **64**, 5 (1862); "Etudes sur la bière." Gauthier-Villars, Paris, 1876.

49. H. Driesch. "Geschichte des Vitalismus." Barth, Leipzig, 1922.

50. E. Haeckel. "Naturliche Schöptungsgeschichte." Reimer, Berlin, 1875.

51. D. Pisaev. "Podvigi evropiskikh avtoritetov." ("Exploits of European Authorities"), Works, Vol. 5. Izd. F. Pavlenkova, St. Petersburg, 1909-1913.

52. H. Bastian. "The Beginnings of Life." Macmillan, New York, 1872. K. Timiryazev. "Vitalizm i nauka" ("Vitalism and Life"), Works, Vol. 5. Sel'khozgiz, Moscow, 1938.

53. M. Wagner. Augsburger "Allgemeine Zeitung," Suppls. 6,7,8, 1874.

54. H. Helmholtz. Preface to the book by W. Thomson and P. Tait. "Handbuch der theoretischen Physik." Braunschweig, 1874.

55. F. Engels "Dialectics of Nature" (Transl. by C. Dutt). Foreign Language Press, Moscow, 1954.

56. H. Richter. *Schmidt's Ihrb. Ges. Med.* **126**, 243 (1865); **148**, 57 (1870).

57. W. Thomson (Lord Kelvin). *Rept. Brit. Assoc.* **41**, 84.

58. H. Helmholtz. "Uber die Entstehung des Planetsystems." Vortrage und Reden, Braunschweig, 1884.

59. Ph. Van Tiegham. "Traité de botanique," Vol. 1. Paris, 1891.

60. St. Meunierref and P. Becquerel. L'astronomie. *Bull. Soc. Astron. France* **38**, 393 (1924).

61. C. Lipman. *Am. Museum Novitates* **588**, 1 (1932).

62. A. Imshenetskii. In the collection "Problemy cvolyutsionnoi i tekhnicheskoi biokhimii" ("Problems of Evolutionary and Technical Biochemistry"), p 20. Izd. "Nauka," 1964.

63. B. Nagy, G. Claus, and D. Hennessy. *Nature* **193**, 1130 (1962); B. Nagy, W. Meinschein, and D. Hennessy. *Ann. N.Y. Acad. Sci.* **180**, 534 (1963).

64. M. Briggs. *Nature* **195**, 1076 (1962).

65. E. Anders. *Enrico Fermi Inst. Preprint* **61**, 51 (1961); W. Titch and E. Anders. *Ann. N.Y. Acad. Sci.* **180**, 495 (1963).

66. S. Arrhenius. "Verldarnas utveckling." Stockholm, 1906; "Das Weltall." Leipzig, 1911; "Das Schicksal der Planeten." Leipzig, 1911.

67. P. Becquerel. *Compt. Rend.* **151**, 86 (1910).

68. E. Graevskii. *Dokl. Akad. Nauk SSSR* **53**, 849 (1946).

69. K. Tsiolkovskii. "Collected Works" Izd. "Nauka," 1964.

70. E. Haeckel. "Generalle Morphologie der Organismen," Vol. 1. Reimer, Berlin, 1886.

71. R. Dubois. *Compt. Rend. Soc. Biol.* **56**, 697 (1904).

72. M. Kuckuck. "Lösung des Problems der Urzengung." Barth, Leipzig, 1907.

73. A. Herrera. *Bull. Labor. Plasmogenie, Mex.* **1**, 49 and 63 (1934); **1**, 71 (1935); *Science* **96**, 14 (1942).

74. S. Leduc. "Les bases physiques de la vie." Paris, 1907; "Theorie physicochim-

ique de la vie." Paris, 1910; *Colloid Chem.* **2**, 59 (1928).
75. L. Lamarck. "Analysis of the Conscious Activity of Man" (Transl. from French); see also A. Studitskii. *Usp. Sovrem. Biol.* **39**, 3 (1955); "Systeme analytique des conaissances positives de l'homme restreinte à celles qui proviennent de l'observation." Paris, 1820.
76. V. Belyaev. "O pervichnom zarozhdenii." ("Primary Generation"). Warsaw Univ. Press, Warsaw, 1893.
77. E. Schafer. Life, its nature, origin and preservation. *Rept. Brit. Assoc.* 3 (1912).
78. K. Timiryazev. From scientific chronicles, 1912. Collected works; *Sel'khozizdat.* **7**, 447 (1939).
79. F. Wohler. *Ann. Phys. Chem.* **12**, 253 (1828).
80. A. Butlerov. *Compt. Rend.* **53**, 145 (1861); *Ann. Chem.* **120**, 295 (1861).
81. E. G'eldt. "Istoriya organicheskoi khimii s drevneishikh vremen do nastoyashchego vremeni"("History of Organic Chemistry from Ancient Times to the Present") (Translated from German).
82. A. Oparin. Doklad na zacedanii Russkogo botanicheskogo obshchestva, 1922 (Reports of the Meeting of the Russian Botanical Society, 1922).
83. A. Oparin. "Proiskhozhdenie zhizni" ("Origin of Life"). Izd. "Moskovskii rabochii," Moscow, 1924.
84. A. Oparin. "Vozniknovenie zhizni na Zemle" ("Origin of Life on Earth"). Biomedgiz, 1936; 2nd ed. Akad. Nauk SSSR, 1941; 3rd ed. Izd. Akad. Nauk SSSR, 1957; Life: its Nature, Origin and Development. Academic Press, New York, 1961.
85. J. Haldane. "The Origin of Life." 1929; "The Inequality of Man." Chatto & Windus, London, 1932; "Science and Human Life." Harper, New York, 1933.
86. H. Urey. "Planets, their Origin and Development." Yale Univ. Press, New Haven, Connecticut, 1952.
87. J. Bernal. "Physical Basis of Life." Routledge & Kegan Paul, London, 1951.
88. S. Miller. *Science* **117**, 528 (1953).
89. T. Pavlovskaya and A. Pasynskii. In the collection "Vozniknovenie Zhizni na Zemle" ("Origin of Life on Earth"). Izd. Akad. Nauk SSSR, Moscow, 1959.
90. *Proc. 1st Intern. Symp. Origin of Life on Earth, Moscow, 1957* (English ed.). Pergamon Press, Oxford, 1959.
91. *Proc. Conf. Origins of Prebiol. Systems, Wakulla Springs, Florida, 1963.* Academic Press, New York, 1964.

CHAPTER 2

BEGINNING STAGES IN THE
EVOLUTION OF CARBON COMPOUNDS

Modern astronomical data show the amazing integration of the entire galactic system into a single entity. The evolution of the Universe is represented to us now as a single process of development, where each successive stage is inseparably connected with the previous one and can be understood only by studying this previous stage. Until recently, in discussing the question of the origin of life, the existence of planets like the Earth with specific dimensions and a specific composition of matter has been somewhat taken for granted. However, now we know that the existence, not only of the Earth, but even many elements necessary for its formation had their source and their history of development in the stellar Universe and can be understood only in the light of a perception of the history of the Universe.

Such perception requires the penetration of the human mind into unimaginable distances of time, a study of events from which we are separated by many billions of years. This became possible only after objective methods were found for the determination of geological and astronomical time in absolute units of measurement—in years, millions or billions of years.

The phenomenon of spontaneous radioactive decay of the heavy elements, invariably proceeding at a constant rate, independent of other external conditions, was first used as a standard for these measurements. The period of decay is proportional to the number of parent atoms. The decay constant of any radioactive isotope thus can be expressed in terms of the "half-life" of the given isotope; that is, the time in which a certain number of its atoms will decay to half their initial value.

The half-lives of different isotopes vary greatly; they are sometimes extremely short, but there are isotopes for which the decay period is measured in billions of years. The half-life of uranium with atomic weight 238, for example, during its conversion to the lead isotope with atomic weight 206 is 4.5 billion years. Determining the relative content of these isotopes in any rock, we can obtain quite an accurate idea of the interval of time since the rock first hardened or crystallized. Other isotopic transformations are also used in addition to the conversion of uranium to lead for measuring geological time; for example, the conversion of thorium-232 to lead-208, potassium-40 to argon-40, rubidium to strontium, etc. (1).

This method is now widely used for determining the age of individual sections of the Earth's crust. We can demonstrate in this way that the time elapsing since the oldest geological shields, located in Karelia, the U.S.A., India, and South Africa hardened lies within limits of about 3.5 billion years. The radioactive age of the oldest rocks of the Kola Peninsula is determined at 3.6 billion years (2). However, such a determination still does not allow us to get a clear estimate of the age of our planet itself, since the Earth's crust must have undergone numerous convective changes and multiple local fusions, which consequently have caused changes in the ratio of the isotopes.

There is no doubt that the Earth as a planet must have been formed considerably earlier than the oldest shields of its crust, since a certain interval of time was required for them to finish congealing (3). At the present time we assume that at least some indication of the Earth's age can be obtained by determination of the radioactive age of meteorites falling on the Earth. These cosmic bodies originate as a result of the disintegration of colliding asteroids (4). Asteroids are extremely small — some hundreds of kilometers in diameter — which precludes the possibility of the gravitational evolution of heat and contributes to rapid loss of heat by radioactive decay. Thus the asteroids must have had a very thermal history in comparison with planets like the Earth.

As a rule, they could not have undergone volcanic fusion; internal convection of matter, transport of silicates, evolution of gases could never have occurred in asteroids. In addition to this the asteroidal matter apparently congealed rapidly immediately after separation of planetary condensations into the newly born solar system from the gas and dust nebula. The radioactive age of meteorites thus is now equated with the age of the planetary matter. Direct radioactive determinations have shown that many of the meteorites congealed 4 to 4.4 billion years ago (5). However, use of the so-called lead method establishes a time of

4.6 billion years since the principal condensation of meteoritic matter. Patterson (6) considers that the age of the Earth can be measured by this method in spite of the thermal blending of rocks and loss of gases. It is interesting that this age was determined by Patterson to be 4.55 billion years, which quite closely corresponds to the age determined by meteoritic investigation. It is true, as Whipple notes (7), that there is also the possibility that the Earth and the parent bodies of the meteorites solidified somewhat earlier.

A detailed study of the individual characteristic isotopes contained in meteorites and the evidence presented above causes us to consider that the formation of the sun and planets from their parent gas and dust nebula was followed from the process leading to the formation of the heaviest elements, including the radioactive elements.

The fact is that a study of the isotopes of uranium undertaken as early as the 1920s enabled Rutherford (8) to arrive at a determination of the time at which the heaviest elements were formed. Rutherford's investigations, and in our time, those of Fowler (9), Baranov (10), and several other authors, have shown that the radioactive elements and also, therefore, the heavy elements which enter into the composition of the Earth and other planets of the solar system came into being not earlier than approximately 5–5.5 billion years ago.

This has been confirmed by the work of Anders (11), Reynolds (12), and others with radioisotopes having a shorter half-life, for example with iodine-129 (half-life 12.7 million years). This work has made it possible to determine the interval of time between the origin of these isotopes and the formation of the Earth. It has proved to be very small, within limits of one to several hundred million years. Consequently, the age of these heavy elements, without which the formation of terrestrial planets would be impossible, does not exceed by far the age of our planetary system; it was formed immediately after these elements appeared in space. However, their origin does not set a limit to the depths of time which we can determine. It is now considered an established fact that star formation occurs uninterruptedly in our galactic system (13). Somewhat conditionally we can assume that our galaxy consists of five different types of stellar populations, differing from one another in chemical composition, distribution of mass, internal motion and absolute age (Fig. 4). The first and oldest type includes stars approximately as old as our galaxy as a whole. The age of our galaxy, on the basis of

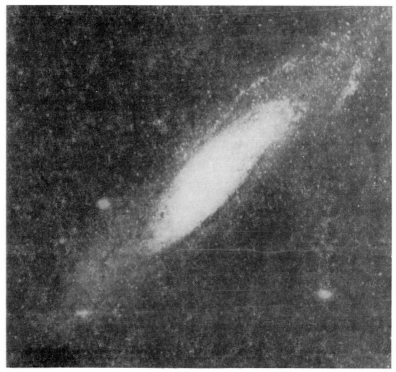

FIG. 4. The nebula in Andromeda, a stellar cluster similar to our galaxy.

some recent data has been determined by Fowler (14) and Hoyle (15) to be 12 to 20 billion years. The oldest stars in our galaxy belong mainly to its central region and are part of the globular stellar clusters. Most of the younger stars are found in the disk-shaped area of the galaxy. Our sun, which is a star of the third magnitude, belongs in this group. Brownlee and Cox (16) have calculated in detail the evolutionary development of our star, and have determined its age at about 5 billion years, which is well within the age limit of our galaxy. Consequently the age of the sun slightly exceeds that of the Earth or of the meteorites as determined by the radioactive method. There are also younger stellar populations in our galaxy; stars in close proximity to the spiral arms of the galaxy belong to this group.

It is very interesting to compare the chemical composition of the oldest and of the relatively young stellar generations. In the stars of the spherical clusters, the oldest objects in the galaxy, hydrogen largely predominates over metals, while the opposite is true in the younger stars, although it would seem that the older a star is, the greater its loss of hydrogen by "burning up." This graphically demonstrates that the primordial medium from which stars of the first generation are formed consists almost exclusively of hydrogen, while stars of later generations, including our sun, are formed from mixtures of light and heavy elements. These latter must therefore have occurred in the galaxy after the first stellar inhabitant had already come into being.

However, terrestrial planets could not have arisen without the heavy elements. A body of a sufficiently large mass, like a star, can be formed from light elements alone, even almost wholly of pure hydrogen. This, in all probability, was what happened in the beginning of our galactic system. This type of mass is completely stable, and when sufficiently compressed, the simple thermonuclear reaction transforming hydrogen into helium and serving as a source of stellar energy will develop. But the existence of bodies with dimensions of our Earth formed entirely from pure hydrogen is absolutely impossible, since the mass of these bodies is not sufficient to prevent dissipation of hydrogen into interstellar space. We may therefore conclude that in the first billion years that our galaxy existed, before the heavy elements had formed in sufficient quantity, a planetary system such as ours could not have existed, and therefore life, which arises during the evolution of such planets, could not exist either. The distribution of life in the Universe must therefore be restricted to star populations of sufficiently late types, made up of all the elements of the periodic table. Thus in studying the cosmic history of the origin of life, we must first of all establish when and how these elements could have been formed within our galaxy (17).

As yet there is no easy answer to this question. In 1959, a special International Symposium devoted to this subject was held at Liege, at which Struve (18) and Cameron (19) reviewed the contemporary state of knowledge concerning the chemical evolution of the stars. Cameron indicated two possible ways in which elements might be formed (nucleosynthesis): a stable process depending on the mechanism of stellar light emission in stable stars; and an unstable process resulting from the explosion of supernovae.

Nucleosynthesis of light elements is exergonic. The energy liberated causes stars to radiate light. Since the electrostatic repulsion when hydrogen nuclei combine is relatively small, the transformation of hydrogen into helium can occur at temperatures on the order of a million degrees. This reaction is possible in the contracting gaseous sphere consisting only of hydrogen, and can begin immediately after this sphere attains the indicated temperature as a consequence of its compression and the freeing of gravitational energy. In contrast to this, in order for the heaviest and radioactive elements to be formed, it is necessary to have a temperature on the order of several billions of degrees, which is lacking in the depths of the stable stars and which is found only in exploding supernovae.

Thus the light emission by stable stars belonging to the main sequence of the Hertzsprung-Russell diagram (Fig. 5) is sustained by the simple nuclear reaction which converts hydrogen into helium, at first through the medium of deuterium and then with carbon as a catalyst. Later as its hydrogen is consumed, the star leaves the curve of the main sequence and is quite rapidly transformed into a red giant. In this way, in addition to carbon, the heavier elements can also be formed from helium in the stellar nucleus (20). However the formation of these elements is limited to the comparatively moderate order of atomic numbers, principally up to neon-20 or magnesium-22, and they cannot even then be thrown off by the star into surrounding space.

Thus Cameron considers that the formation of all the elements in the periodic table, including the heavy radioactive elements, was possible only during explosions of supernovae. These catastrophic stellar explosions in the present epoch occur approximately every 300 years in our galaxy. They produce billion degree temperatures and monstrous, though short-lived, pressures. Enormous amounts of energy are thereby released and a colossal ejection of matter occurs during which the star loses in a short time a mass comparable with the mass of the sun. The relatively light elements present in the stellar matter quickly increase their atomic weights by capturing the neutrons escaping from the stellar nucleus. The heavier elements of the periodic table originate in this way. Part of the stellar matter is dispersed by shock waves in all directions and thus enriches the surrounding space with heavy elements.

Fesenkov (21), developing this theory, concluded after thoroughly investigating the isotopic composition of meteoritic matter that the sun

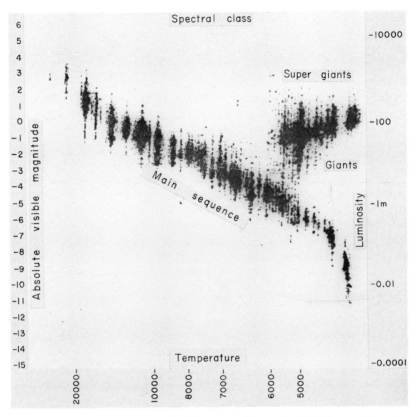

FIG. 5. Distribution of stars of various spectral classes in the Hertzsprung-Russell diagram.

and its surrounding planets were formed in a single process associated with the explosion of a supernova in the corresponding region of the galaxy. This explosion occurred approximately 5 billion years ago. It furnished heavy elements to the solar system being formed, enriching it in this respect in comparison with stars formed earlier than the sun. In addition to this, the supernova explosion was accompanied by widely disseminated shock waves which could have caused the compression of interstellar matter into a proto-solar gas and dust cloud, imparting gravitational unstability to it and serving as an impetus for the subsequent formation of the sun and planets. A similar type of supernova

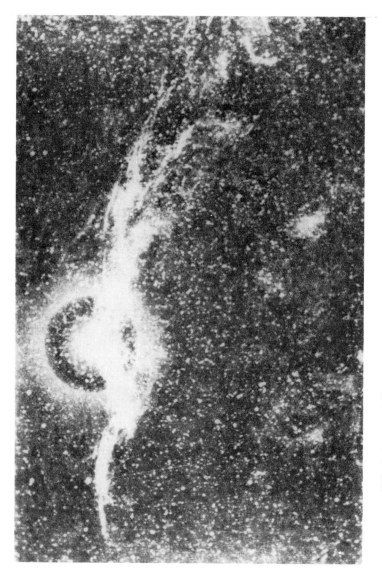

FIG. 6. System of filamentary nebulae in the region of the constellation of the Swan.

explosion must have occurred many times in our galaxy both before and after the emergence of the solar system. As an example of the result of similar events, Fesenkov points out the region of the constellation of the Swan, occupied by a whole system of filamentary nebulae (Fig. 6). An investigation of various photographs of this region, made at intervals of 50 years, indicates that these filaments are being dispersed at a significant rate from some general center which is a source of radio emission. Such a phenomenon has interested Fesenkov as a demonstration of an explosion of a supernova in the past in this region of the galaxy.

The Mountain Astrophysical Observatory of the Astrophysical Institute of the Kazakh SSR Academy of Sciences also discovered among these filamentary nebulae, a number of stellar chains – unstable, recently formed, each consisting of stars resembling each other in magnitude and color and in the same stage of evolution. "In a comparatively short time"–writes V. Fesenkov, "these unstable stellar chains will cease to exist as such. The stars composing them will be dispersed, will lose all relation to each other and will pass through independent but similar stages of evolution." Thus Fesenkov associates the origin of the solar system with the explosion of a supernova occurring not long before the formation of this system; the explosion caused a local condensation in the surrounding gas–dust medium, resulting in star formation, and simultaneously established the nucleosynthesis of heavy elements necessary for the formation of the terrestrial planets.

Frank-Kamenetskii conceives the route of nuclear transformation occurring in the formation of our solar system from another viewpoint (22). This author for several years has indicated the possible significance of "cold" processes in the formation of elements. In this case, the nuclear particles obtain the energy which they need for nucleosynthesis, not from heating, for which "million degree" and "billion degree" temperatures are necessary, but from the direct acceleration in shock waves and electromagnetic fields.

At the time of creation of the solar system, our sun was in the stage of gravitational condensation, in the stage of congealing from a gas–dust cloud. It was still a very young star, and according to Frank-Kamenetskii, it then would belong to young stars of the Tauri type which have been observed at the present time. These stars are in a state of violent activity. An unusual nonthermal radiation source is found in their spectra; this is direct evidence of the fact that here electrons obtain their energy not from heating but from some sort of "cold" process. This indicates the possibility of a similar type of acceleration in the nuclear particles. Thus the violent motion in the plasma of young stars, causing turbulent waves and variable magnetic fields to form, brings about conditions necessary for a "cold" synthesis of elements.

Comparing data obtained by a study of the isotope content of several elements (specifically lithium, beryllium, and boron) in the earth's crust, meteorites, the sun and young stars, Frank-Kamenetskii concluded that the birth of the planetary system is achieved just at the moment when the star passes through the stage of the T Tauri type in its evolution. "Observing the violent transitional phenomena of the non-thermal luminosity of these stars" – he writes, "we are seeing the 'birth pangs' of the planetary system emerging into the light."

Be that as it may, we assume that our planetary system was formed from proto-planetary matter which included all the elements of the periodic table and which originated not long before or even at the same time as the formation of the planetary system itself. Later only a few changes in the isotopic composition could have taken place due to the influence of cosmic radiation, but these changes are insignificant when compared with the processes of formation of the elements which were described above.

It is usually assumed that the earth, like other planets of our system, was formed of matter from the disk-shaped gas and dust cloud, which at some time surrounded our sun (23). It may be that this cloud was captured by the sun out of cosmic space (24), but most likely it was a residue of that gas and dust nebula from which our sun itself was formed (25). In any case, its chemical composition can be compared with the composition of present-day clusters of interstellar gas and dust. The gas in this case is mainly hydrogen, which is the element predominating in the cosmos. Helium and other inert gases are present in lesser quantities. The remaining elements, particularly the heavy metals, constitute only hundredths, thousandths, and even smaller fractions of a percent of the overall mass of the cluster. This composition is directly comparable to the overall composition of the present-day sun, which must contain an overwhelming proportion of the matter in the entire solar system.

It contains 60% hydrogen, 30% helium, 1-2% carbon, nitrogen, and oxygen and a total of less than 1% of heavy elements (26).

According to Brown (27), all the matter occurring in the original gas–dust solar nebula can be classified into the following three groups. To the so-called gaseous group belong hydrogen, helium, and the noble gases which remain in a gaseous state at temperatures close to absolute zero. At the very low temperatures prevailing in the original nebula they have comprised the chief gaseous component. Next is the group termed

TABLE 1

Group	"Earthy"	"Icy"	Gaseous
Elements	Si, Mg, Fe and others (plus O)	C, N, O (plus H)	H, He
Relative Mass	1	4–7	300–600
Melting Point	2000°K	≤273°K	≤14°K
Planets			
Terrestrial planets	1.00	<0.01	0
Jupiter	<0.01	0.1	0.9
Saturn	0.01	0.3	0.7
Uranus	0.1	0.8	0.1
Neptune	0.2	0.7	0.1
Comets	0.15	0.85	0

the "icy" group, which is composed of matter formed from carbon, nitrogen and oxygen (plus hydrogen) such as ammonia, methane, and water. Finally there is a group conditionally called the "earthy" group, which consists of compounds of silicon, magnesium, iron, etc.; its melting point is very high. The proportions of these groups in the sun's composition are given in Table 1 (28).

However, when the planets were formed from the disk-shaped proto-planetary cloud, a far-reaching redistribution of this matter occurred, evident from the data of Table 1, in which present-day proportions of these groups are given for various planets of the solar system.

For what reasons and by what processes this redistribution of matter occurred when the individual planets were formed, we can only theorize as yet. At the present time there are a great many hypotheses in the cosmogonic literature concerning the mechanism of formation of the solar system (29). Both the turbulent motion in the gas–dust medium and gravitational forces, and at a later date, in particular, magnetohydro-dynamic phenomena (30) have been enlisted for this purpose. The majority of contemporary cosmogonists starting from general physical laws have striven by means of corresponding mathematical computations to explain rationally the way in which these structural features of our solar system were formed: the regularities in the motion of the planets and their satellites, the form of their orbits, the planetary distances, the dimensions and mass of the planets, the distribution of angular momentum between the sun and the planets, etc. But it is doubtful whether any of the hypotheses now existing can completely satisfy all the demands required of it.

Whipple (7), in a report which he recently gave at a meeting devoted to the 100th anniversary of the National Academy of Sciences in Washington, assumed as the starting point for the formation of the solar system the development of a cloud of gas and dust which was a segment of the gigantic interstellar clusters. It has dimensions commensurable with the present-day solar system, a low temprature of the order of $50°K$ (about $220°$ below $0°C$), a weak magnetic field, and it contained approximately 1000 hydrogen atoms per cubic centimeters, as well as a considerable amount of dust. It was gravitationally unstable and exhibited a certain turbulence and some central condensation that in the final result led to its general collapse. When the mass of the central body increased precipitately it was rapidly condensed and heated by the gravitational energy released, and then by the hydrogen-helium reaction. The central mass was thus transformed into our star and the remainder of the matter, not becoming part of the sun, formed a disk-shaped cloud; this was the beginning of the formation of "protoplanets."

In this protoplantetary cloud there must have inevitably arisen a process of gradual unification of the dispersed matter, at first into comparatively small bodies, and then to larger structures. The temperature of the cloud largely depended, on the one hand on solar radiation, and on the other, on heat loss by the cloud through light emission into interstellar space. Thus there was a tremendous temperature gradient between the periphery and the central regions of the gas and dust disk. This circumstance also must have determined the distribution of protoplanetary matter in the disk.

Close to the periphery, where intensive cooling had occurred, the "icy" group elements condensed rapidly and, "froze" on the "earthy" particles, forming the so-called cometesimals. In the region of the orbit of Jupiter and Saturn, gas, dust and cometesimals were combined rapidly into the giant planets, to which the magnetic lines of force of the sun contributed. Still farther out toward the periphery, cometesimals were collected into the planets Uranus and Neptune, which were mainly formed from cometesimal matter. Consequently part of this matter entered into the composition of Jupiter and Saturn, part formed Uranus and Neptune, part fell on the sun or remained circling about it in an exceptionally elongated orbit and was transformed into comets. Finally, a certain amount of cometesimal matter was lost entirely from our planetary system by flying off into interstellar space.

Planet formation in regions of the discoid cloud close to the sun, where the terrestrial planets were formed, proceeded differently (31).

Here the gaseous component of the cloud was almost wholly lost, "ice" was only partially retained and "earthy" group compounds predominated. Accumulation of dust particles led to the formation of the so-called planetesimals, small bodies incorporating all the nonvolatile substances of the primordial dust cloud — silicates and their hydrates; metals (iron in particular) and their oxides, sulfides and carbides; as well as a small amount of the "icy" group components. Further consolidation of the planetesimals led to the formation of the terrestrial planets and the asteroids.

The planets originating in this way had then a long and quite complex thermal history while the asteroids kept their initial composition and structure more or less unchanged. However, considerable heating occurred in the center of some of them, due to the decay of the "short-lived" radioactive isotopes. The iron-nickel cores were thereby formed, which when the asteroids were split up by collision, originated the iron meteorites. The stony meteorites, however, were splinters from the periphery, or they were formed when the asteroids which had kept their initial composition and structure were split open.

Thus, according to the hypothesis now accepted, our planet was formed by the accumulation of cold solid bodies, heterogeneous in composition, with a different iron and silicate content, but in general lacking such free gases and volatile compounds as molecular hydrogen, volatile inert elements, or methane (32). Because of the nearness of the region where the Earth was being formed to the Sun, these gases must have forsaken this area and collected in the colder region of the proto-planetary cloud, for example where Jupiter or Saturn was being formed. Only a small quantity of these gases, adsorbed by solid rocks, were retained by the Earth, which had just been formed, and became part of its primary atmosphere. However, this atmosphere could not last very long, since the gases composing it were not restrained by terrestrial attraction and were dissipated into interplanetary space. In this manner, the Earth, having lost its primary atmosphere, including several noble gases widely distributed in the Universe, acquired as a whole its present-day mass and composition, which is very close to the average composition of present-day meteorites (33). The age of matter on the Earth, originating in this way, can be fixed at approximately 4.5 billion years.

The compact mass of the Earth continued to evolve due to its internal energy, and was divided into a nucleus, mantle, and crust. This event can be dated at a time more than 4 billion years ago. The mechanism of this differentiation of the primarily homogeneous terrestrial matter still remains unclear. The explanation often has been advanced of a secondary melting of the planet due to the initial gravitational heat and radiogenic warming. However, a period on the order of several billion years would have been required for complete melting of the Earth, and actually differentiation could have occurred within an incomparably shorter period. Perhaps the "short-lived" and at present "extinct" isotopes such as ^{10}Be, ^{26}Al, ^{129}I, ^{244}Pu, etc. played an important role in the complete or partial melting of the Earth.

In any case, all further progress in the evolution of our planet was determined by its thermal history (34). At present the climatic conditions on the surface of the Earth depend mainly on the intensity of solar radiation and the heat emission of the irradiated surface. When our planet was formed, the basic sources of its internal heat were the gravitational energy given off by the agglomeration and condensation of terrestrial matter and energy resulting from the breakdown of radioactive elements, which easily can be calculated by the law of radioactive decay.

The most important limits for our further discussion on the development of the Earth are the problems relating to the formation on its surface of a crust, an atmosphere, and a hydrosphere.

The present-day crust of the Earth consists of granitic and basaltic envelopes overlaid with a cover of sedimentary rocks. Under the crust lies the so-called mantle, characterized by ultrabasic rocks poor in silicon (dunites). Vinogradov (35), on the basis of his experiments on zone melting of stony meteorites (chondrites), became convinced that the dunites of the mantle were residues of the melting of the original terrestrial matter, similar in composition to the chondrites. Melting out of light basaltic rocks from the strata of the mantle under the influence of radiogenic heat must have been accompanied by the evolution of various vapors and gases as they were given off when the temperature increased or as they formed in the solid earth envelopes from the radioactive, radiochemical, and chemical processes being accomplished there.

Thus the formation of the water and gas envelopes of the Earth (its hydrosphere and atmosphere) were from the very beginning clearly connected with processes occurring in the lithosphere.

The amount of water on the Earth's surface at its beginning must have been considerably less than at the present time. According to Urey, the primordial Earth had only about 10% of the water of present-day seas and oceans. The remaining water appeared later, gradually being formed from silaceous hydrates or from the bound constitutional water of the Earth's interior when the lithosphere was formed.

The age of the oceans is approximately evaluated according to the content and rate of entry of certain chemical components, for example sodium or phosphorus. These values indicate that the age of the ocean is approximately 3 billion years or more.

The age of the secondary atmosphere formed by the gases evolving from the lithosphere can be determined by the content of argon-40 which was formed on radioactive decay of the isotope potassium-40. A value within the limits of 3.5–4.0 billion years was obtained (36). In spite of the fact that the Earth when it was formed was deprived of most of its share of proto-planetary hydrogen, its abundance in the original material had a significant effect on the subsequent composition of our planet and in particular its secondary atmosphere. Even after loss of hydrogen as a free gas, the Earth must have retained a very large amount of this element in the form of various compounds, many of which were given off during formation of the Earth's crust, thus providing an incipient atmosphere and giving it a decidely reducing character. In addition to water vapor, ammonia, hydrogen sulfide and gaseous hydrocarbons were the gases present.

The molecular oxygen characteristic of the present atmosphere was almost entirely absent. It is true that under the influence of short-wave ultraviolet light, some decomposition of water vapor can occur, resulting in the volatilization of hydrogen and abiogenic formation of free oxygen (37). However, even in this case oxygen cannot be accumulated in the atmosphere in any sort of significant quantities, since it is very rapidly and completely adsorbed into rock formations unsaturated with respect to it (38).

We can observe a similar process of oxygen adsorption even now. For example, the lava masses ejected onto the Earth's surface by modern volcanic eruptions usually are very rich in the lower oxides of metals. When they come into contact with atmospheric oxygen they begin to oxidize intensively, to catch fire, as it were, reaching a higher temperature than they had in the depths of the earth, emitting light, melting, etc.

The generally known fact that igneous plutonic rocks (lavas and basalts) have black, gray, and green colors, and the sedimentary formations formed from them (sand, clays) are colored yellow and red is the result of the transformation of their iron content from the ferrous oxide to ferric. Thus, even now, the abundant free oxygen in the atmosphere is sufficient only to oxidize the surface layer of the Earth's crust. If now all the green plant life continually replenishing the content of atmospheric oxygen were to disappear, free oxygen would disappear in the course of about 3–4000 years, having been adsorbed by rocks unsaturated with respect to oxygen (39).

Before the origin of green vegetation, free oxygen that was formed by abiogenic means was so small that the rock formations (including the sedimentary rocks) could not in any measure be saturated with oxygen, and there was in fact no oxygen in the atmosphere. This secondary reducing atmosphere lacking molecular oxygen existed for quite a long time. The date can be established on the basis of direct geological evidence obtained by a study of ancient Precambrian strata.

Rutten (40), using these data, divides the entire existence of our planet into two basic periods – the actualistic and the pre-actualistic. Geologic processes occurring on the Earth's surface during these periods differed widely. As an example can be cited the weathering of rock formations in the following basic steps: (1) erosion of the original rocks, (2) transport of the products originating as a result of the erosion, and (3) their deposition in formations of sedimentary rocks. In the actualistic epoch, in which the atmosphere contains abundant free oxygen, chemical processes play the principal role in the weathering of basic rock formations. Minerals already oxidized are transported and deposited. On the contrary, the atmosphere in the pre-actualistic epoch was a reducing one and thus such minerals as feldspar, sulfites, etc. were chemically stable. They were reprocessed only physically when they were transported and they were deposited in a chemically unchanged form. This is shown by the fundamental work of Rankama (41), investigating ancient deposits of detrital rocks around the granite cliffs of Finland, the data of Ramdohr (42) studying the gold-bearing seams of old shields in South Africa, Brazil, and Canada, the data of Lepp and Goldich (43) on ancient deposits of iron ore in North America, Lorraine, and Luxembourg, and many others.

After a study of these data, Rutten concluded that all these formations could have arisen only through a lack of free oxygen in a reducing atmosphere and in order for this to be true, it must have been present on the Earth's surface at a period at least 2 billion years before our time.

Sagan (44) has postulated that signs of an oxidizing atmosphere appeared considerably earlier. Rutten, however, on the basis of objective geological data, shows that a similar modern oxygen-rich atmosphere was formed only within the total limit of 1 billion years ago. According to Rutten, between the dates which he gave (2 and 1 billion years) lay a transition epoch during which the Earth's atmosphere was gradually enriched with free oxygen, and during which life was developing; life had appeared in the pre-actualistic, as is shown by comparative biochemical and paleontological data.

We can sum up what has been said so far in the following chronological table of astronomical events which led to the formation of our planet (Table 2).

How in the background of these events did the evolution of carbon compounds occur? This is the narrow specific branch of the development of matter which especially interests us, since it is the pathway for the origin of our life on Earth, and perhaps also of life on other similar heavenly bodies.

As we have seen above, carbon had already originated before the formation of the heavy elements, independently of the explosion of the supernovae, in the stable process of stellar light emission. Thus it is an element very widely dispersed in the cosmos (45). It is found in the spectra of all classes of stars, particularly in the oldest populations. In stellar atmospheres having the highest temperatures carbon is present in an ionized or neutral atomic state (46).

It is still impossible at this stage for chemical compounds to form.

TABLE 2
Chronology of Astronomical and Geological Evolution

Event	Time in billions of years before our time
Origin of the oldest stars in our galaxy	12–20
Origin of heavy elements	5.3
Origin of the sun and planets of our system	5
Formation of the Earth with its present–day mass and composition	>4.5
Differentiation of Earth materials, formation of the Earth's crust	>4
Formation of the oldest minerals presently known	3.6
Formation of the oceans	>3
Formation of the secondary reducing atmosphere	>3.5
Geologically determined limit of the existence of a reducing atmosphere	2
Formation of an oxygen atmosphere of present–day composition	1

But as soon as conditions become favorable, carbon immediately combines with hydrogen; this combination can be postulated a priori because of the almost quantitative predominance of hydrogen in the cosmos. And actually, compounds of carbon and hydrogen can be discovered on the most diverse heavenly objects at very different temperature and gravitational conditions.

Signs of these compounds can even be found in the spectra of type A stars (47). In spectra of stars of succeeding types, the hydrocarbon bands appear with more and more clarity as the temperature of the stellar surface decreases and they attain maximum clarity in the spectra of M and R type stars. Our sun, as is known, is a type G star (yellow star). The temperature of its atmosphere is about 6000°C. Spectroscopic analysis shows that a compound of carbon and hydrogen in the form of methane (CH) is present on the sun, and perhaps more complex structures also, containing several atoms of carbon and hydrogen (48). A compound of carbon and nitrogen, cyanogen (CN), is also found on the sun.

Thus we see that hydrocarbons are widely distributed on the surface of stars with temperatures of several thousands of degrees and a very strong gravitational field. On the other hand, we also find them at the opposite extremes in the interstellar gas and dust, at an extremely low gravitational gradient, and at temperatures close to absolute zero. The formation of the simplest hydrocarbon radicals, CH and CH+ have been observed by the workers Kramers and ter Haar (49). According to Urey, the free radicals in the gas and dust aggregates must be transformed into stable hydrocarbons by the catalytic action of dust and the abundant hydrogen. Actually, at the present time, it has been possible to demonstrate directly the presence of methane in the cosmic gas and dust aggregates (50), which are the prototype of that matter from which our solar system was formed.

The formation of the simplest carbon compounds at a considerable distance from the stars under low temperature conditions is also confirmed by data on the chemical constitution of comets (51). These cosmic bodies are formed under conditions almost equivalent to those in interstellar space (52), in the neighborhood of the orbit of Pluto, and from time to time they penetrate the inner regions of our planetary system where their composition may easily be verified by spectroscopic analysis. These investigations have shown that comets abound in light hydrocarbons, cyanogen, and carbon monoxide.

Within the boundaries of the solar system itself we can find enormous amounts of methane in the atmospheres of the giant planets — Jupiter, Saturn, Uranus, and Neptune (53). It is very interesting that methane is present in the atmosphere of Titan (a satellite of Saturn) (54). Titan is 1/40 the mass of the Earth, but it can retain its methane atmosphere because of the very low temperatures ($-180°C$) prevailing in its area.

A study of the hydrocarbon compounds of meteorites is of especially great significance for the question under discussion, first because meteorites are as yet the only nonterrestrial objects which we can investigate directly both chemically and mineralogically, and second because of the similarity of meteorite material to that from which the Earth was formed (55).

The problem of the origin of meteorites has undergone lively discussion in modern scientific literature. It is considered as established that meteorites are formed in our solar system mainly between the orbits of Mars and Jupiter in the zone of the so-called minor planets — asteroids which apparently are "maternal bodies" of the meteorites.

A study of the various radioactive isotopes present in meteorites permits us to date individual events in their history. The age of the heavy elements composing them is the same as that of other bodies in the solar system (about 5 billion years). The time elapsing since the hardening of meteorite material is calculated on the basis of a study of lead isotopes, rubidium and strontium, rhenium and osmium, as being 4.5 billion years. The magnitude of the loss of several volatile radiogenic isotopes (helium, argon) demonstrates the fact that the matter in some meteorites at some period of time could have been at a comparatively high temperature. The time when small fragments were formed from the large asteroids (the so-called cosmic age of the meteorites) has been established by the determination of isotopes formed by the action of cosmic radiation at only tens or hundreds of millions of years. The regularity in the composition and internal structure of meteorites demonstrates the various conditions under which individual groups of meteorites differing in their chemical composition were formed (56).

Usually all meteorites are divided into two basic groups – the stony meteorites and the iron meteorites. Iron meteorites consist mainly of nickel-iron (90% Fe, 8% Ni, 0.5% Co) and contain small amounts of phosphorus, sulfur, copper, and chromium. The carbon in them is mainly in the native form, or in compounds with metals [cohenite – $(FeNiCo)_3C$]. The structural features of these meteorites demonstrate that nickel-iron first occurred in the molten state at 1500°C, then during very slow cooling in the course of tens of millions of years the alloy crystallized and recrystallized. Apparently these meteorites are fragments of the central core of the "parent bodies." The stony meteorites, which fall more frequently on the Earth, come from the peripheral areas of these bodies. Their iron content is considerably lower, and silicates and oxides of such metals as magnesium, aluminum, calcium, sodium, etc., predominate. Most stony meteorites contain not more than 0.1–0.2% carbon. But there are two types of meteorites which contain a very high amount of carbon, varying within limits of 0.45 to 4.8%. These are the carbonaceous diamond-bearing achondrites (ureilites) and the carbonaceous chondrites (Fig. 7). There are very few of the latter in world meteorite collections, but they have special interest for us, since the presence of certain minerals in them demonstrates that they never were heated higher than 300°C (57).

Carbonaceous chondrites are quite brittle structures consisting of a black opaque mass in which small prophyritic chondrules and granules of olivine and pyroxene are embedded. The carbonaceous chondrites are characterized not only by a high carbon content but also by water bound in the minerals (hydrated silicates and alumino-silicates). The

FIG. 7. The Mighei carbonaceous chondrite. Weight about 1600 gm. (Collection of the Commission on Meteorites of the Academy of Sciences of the USSR.)

carbon in carbonaceous chondrites consists of amorphous carbonaceous matter and graphite. It can also be in the form of organic compounds and carbonate. The adsorbed gases consist of CO_2, CH_4, CO, and H_2 as well as Ar, Ne, and Xe. The organic compounds are mainly hydrocarbons, partially oxidized and sulfur-containing. In addition to the low molecular weight hydrocarbons, the carbonaceous chondrites also contain polymerized compounds, mainly of the aliphatic series (58). There are indications in the scientific literature on the discovery in meteorites even of such complex compounds as pyridine bases or amino acids (59). However, these indications still require further experimental verification (60). Mueller (61), subjecting to detailed analysis all facts known at the present time concerning the chemistry and petrography of both inorganic and organic phases of carbonaceous meteorites, and also weighing all the data about their occurrence, came to the conclusion that the carbonaceous compounds of these meteorites undoubtedly arose in the process of chemical evolution entirely independently of the life process. Abiogenic origin of meteoritic organics has been confirmed also by the recent work of Briggs (62) who studied polymeric organic subtances of carbonaceous chondrites. Such investigations are of great interest since they open before us a route to the study of the abiogenic formation of organic substances under natural conditions, somewhat different, it is true, from those which existed on Earth before the emergence of life.

Later we will return to the many broad investigations made in our time of these meteoritic materials when we discuss the means by which organic polymers were formed in the evolution of cosmic bodies (see Chapter 3). Here we want only to emphasize the fact that we can prove the existence of carbon compounds mainly in the reduced form as hydrocarbons on the most diverse objects of our stellar world – in stars, in gas and dust clouds, in planets and their satellites, in comets and meteorites.

Most of the heavenly objects listed above have developed and are developing by other means than is our planet; however as we have seen, the origin of the most primitive carbon compounds and the beginning stages of their evolution are almost common in the Universe surrounding us. The fact that the Earth is not an important exception in this respect can be verified on the basis of knowledge of the evolution of carbon compounds during the formation of our planet.

As we have shown above, long in advance of the formation of the Earth as a planet from the protoplanetary discoid cloud, the gaseous components of this cloud (hydrogen and the noble gases) must generally have been volatilized and substances of the "icy" group must also have partially disappeared. This is particularly characteristic of methane, which retains its gaseous state even at very low temperatures.

Aston (63), 40 years ago, first focused attention on the low content of noble gases in the Earth's composition and explained this by their chemical inertness, their incapability of combining with other elements into heavier molecules. Later Suess (64) and Brown (65) indicated that the content of noble gases on the Earth was very much less than in the cosmos. They were "depleted" to a large extent as a result of their volatilization from that part of the protoplanetary cloud in which the

TABLE 3
"Depletion Factor" of Volatile Elements and Compounds on the Earth

Elements and Compounds	Relative content in the cosmos $Si = 10,000$	Moles/cm^2 on Earth	"Depletion" in comparison with xenon	Proportion remaining on Earth
Ne	116,000	6.5×10^{-4}	0.9×10^{-4}	0.9×10^{-11}
Ar^{36+38}	1,120	1.2×10^{-3}	2.6×10^{-2}	2.6×10^{-9}
Kr	0.718	3.5×10^{-5}	1	1.0×10^{-7}
Xe	0.0665	2.8×10^{-6}	1	1.0×10^{-7}
H_2O	111,000	15,000	3,200	3.0×10^{-4}
C	40,000	350	210	2.0×10^{-5}
N	80,000	54	16	1.6×10^{-6}

Earth was formed. Urey (66) gives a table of data on the relative content in the cosmos and on Earth of some elements and their compounds. In Table 3 the value of the "depletion factor" relative to xenon is pointed out, as well as that proportion of the gas or element retained on Earth of its total content in the gas and dust cloud.

It is evident from the data of Table 3 that the amounts of volatile elements and compounds are considerably "depleted" in comparison with the nonvolatile elements. Since this division was accomplished even before the Earth condensed into a solid body, it occurred in a low gravitational field. It is easy to understand why the neon content is "depleted" to a considerably greater degree than xenon, but the water compounds of C and N are abnormally high from this point of view. If "depletion" took place wholly by means of fractionation according to mass, that is, if compounds were volatilized from the gravitational field according to molecular weight, then CH_4 should have been "depleted" to the same degree as neon. In addition to this, if "depletion" depended on the vapor pressure, solubility, adsorption, or formation of hydrates, then the "depletion factor" for CH_4 would have had to be between the values for argon and krypton, but in fact a tremendously large fraction of the carbon was retained on the Earth.

The only acceptable explanation for such a large carbon content on the Earth is the theory that it was retained in the form of some sort of heavier chemical compound than methane when our planet was formed. Undoubtedly such compounds are the metal carbides, graphite, or amorphous carbon found in meteorites; however, organic substances to some extent could also have been influential in retaining on the Earth a significant part of the original carbon. The finding of these organic compounds in carbonaceous chondrites is an indication of this.

In this connection many authors at the present time have expressed the opinion that a considerable amount of organic substances must already have been synthesized before the formation of the Earth as a planet. Thus, for example, Lederberg and Dowier (67) consider that organic compounds were formed by the effect of corpuscular radiation on the particles of maternal cosmic dust. Fowler, *et al.* (68) also emphasize the great significance of the formation of organic compounds in that dust cloud from which the Earth was formed. Glasel (69) in model experiments has synthesized acetylene, ethane, propane and some other hydrocarbons by bombardment of CH_4 with electrons at a low temperature. R. Berger showed that urea, acetamide, and acetone are synthesized from CH_4, NH_3, and H_2O at $77°K$ ($-196°C$) as a result of the action of high-energy protons. Very many such examples can be cited.

Recent broad investigations of the organic matter of the carbonaceous chondrites as well as some data on the nature of "noctilucent clouds" and the meteoric dust of red clays from ocean bottoms has stimulated Bernal (70) to form his new hypothesis of the cosmic origin of these organic substances which served as material for the formation of living matter on the earth. According to this hypothesis, the synthesis of more and more complicated organic molecules and their polymers occurred even in the particles of cosmic dust covered with ice, and by condensed gases, under the influence of solar radiation and the action of cosmic rays. As the temperature increased in the zone of formation of the terrestrial planets, the frozen and condensed layers on the surface fumed and volatilized, but the more or less high molecular weight organic substances were retained. This, then was a beginning for the formation of those carbon complexes which can now be found in carbonaceous meteorites. The Earth, being formed by the accumulation of chondritic bodies at a relatively low temperature, obtained from them a ready-made complex of organic materials part of which could have appeared on the planetary surface in a relatively unchanged form and would have served as a basis for the subsequent development of life. The possibility of such a synthesis of organic matter on particles of cosmic dust finds its confirmation in the series of facts stated above.

However, Miller and Urey (71) contend that organic matter of a more or less complex composition, even if it were synthesized on particles of "mother substance" before the Earth was condensed into a solid body, could not play a decisive role in the emergence of life on the surface of our planet. When the Earth was formed from planetesimals the organic matter must have been more or less equally distributed over its entire mass and its escape to the surface in unchanged form would be quite unlikely. Even if our planet was not completely melted, its internal temperature because of gravitational energy and radioactive heating would be fully sufficient for the pyrolysis of organic compounds in its interior. Only products of this pyrolysis such as CH_4, CO_2, CO, N_2, NH_3, H_2O and H_2 could be evaporated from the Earth's surface and enter into the composition of its secondary atmosphere, since now most of these gases (with the exception of H_2) were already held down by the terrestrial attraction of the solid planet being formed. A considerable part of the carbon dioxide (CO_2) of the secondary atmosphere, combining with silicates of the lithosphere, was precipitated on the Earth's crust in the form of carbonates, deposits of which arose long before the appearance of life. The reduced forms of carbon and nitrogen (CH_4 and NH_3) served as material for the secondary formation of all complex organic substances.

However, as we have indicated above, the carbon compounds retained on formation of the Earth may not only have been organic substances far from being more complex than methane, but also primarily amorphous carbon, graphite, and metal carbides. These latter, interacting with hydrated rocks, must have produced gaseous hydrocarbons on a wide scale on the Earth's surface. Native carbon interacting with hydrogen from radiochemical breakdown of water was also a source of hydrocarbons when the Earth's crust was formed. Thus the generation of hydrocarbons and their close derivatives on the Earth's surface as the lithosphere originated not only could be, but had to be accomplished in numerous ways: (1) as a result of the reaction of carbides with the hydrated rocks; (2) direct reduction of graphite and amorphous carbon by free hydrogen, formed during the radiochemical breakdown of water; (3) pyrolysis of the primary organic matter; (4) generation of the primary methane absorbed by rock formations, etc.

Now, the only argument can be which one of the possible pathways of abiogenic formation of hydrocarbons generally predominated during the period of the Earth's formation and in the succeeding epochs. But the fact itself of abiogenic formation cannot presently be doubted. It is confirmed in a series of recently discovered geological findings.

Begun during the earlier periods of our planet's existence, the formation of its crust is not complete even at present. As a result of this, the abiogenic formation of hydrocarbons occurs even now at several points on the Earth's sphere (72). This is particularly evident in a number of gas deposits having no direct connection with sedimentary rocks. Such, for example are the hydrocarbon gases formed in the crystalline rock formations on Lake Huron in Canada or the Ukhtinsk deposit in the U.S.S.R., where hydrocarbons in very large amounts can be found in fissures in igneous rocks. According to a communication of Sokolov (73), he discovered that methane, ethane, propane, and even higher molecular weight hydrocarbons were present in igneous rocks at a number of locations in the Soviet Union.

Of considerable interest in this respect are the gas emissions found recently in the Khibinsk massif on the Kola peninsula. Their analysis shows methane and heavier hydrocarbons to be present. A very careful examination of the physical-chemical and geological situations existing here definitely suggests a magmatic origin of these gases (74).

The number of oil deposits found in igneous and metamorphic rocks steadily grows, although comparatively little attention has been paid to them, in view of the fact that they rarely have industrial significance. Nevertheless they now are numbered in the hundreds (75). In particular, the liquid and gaseous hydrocarbons in the form of smears and insignificant deposits were also found by deep boring along fissures in metamorphic and crystalline rock formations of the bed where the hydrocarbons scarcely could penetrate from sedimentary rocks.

Thus although oil extracted from sedimentary deposits clearly bears traces of its biogenic origin, we must not in the light of facts presently known deny the fact that abiogenic formation of hydrocarbons can occur even in the modern epoch.

Robinson (76), studying in detail the various specimens of oil obtained from numerous deposits, turned his attention to the fact that crude oil of more ancient origin contains substances very far removed from biochemical compounds. The origin of these substances from biological matter does not yield to a rational explanation acceptable to chemical

and geological points of view. On the contrary, their abiogenic formation is easily understood and experimentally reproduced under conditions close to nature. Robinson therefore concludes that a two-fold
origin of petroleum, both abiogenic and biological, is necessary. He
convincingly demonstrates that the more ancient the petroleum studied,
the more it shows signs of its abiogenic origin.

We can hardly now doubt that during the initial period of the Earth's
existence there was present in the atmosphere and lithosphere on its
surface a considerable quantity of hydrocarbons which were the original
material for the subsequent evolution of organic compounds.

The main mass of these hydrocarbons originating when the Earth's
crust was formed had, as we have said, an endogenous origin. However,
at a later date more work has appeared showing that during the whole
course of existence of our planet it has also been "fattened up" by
organic matter reaching it from outside, i.e., exogenously. The meteorites falling on the earth, particularly the carbonaceous chondrites were
of especially great significance in this respect (77). Their organic matter
must not have been subjected to strong heating and pyrolysis and this
means that they could have been directly influential in the subsequent
synthesis of complex organics occurring on the surface of the Earth.
Recently Oro (78) has advanced the theory that in addition to organic
matter of geochemical origin on the Earth, there was a considerable
proportion of hydrocarbons borne to our planet not only by meteorites
but also by comets.

Urey (79) has calculated that during the existence of our planet not
less than 100 direct collisions of the Earth with comets must have
occurred; the matter from these foreign bodies remained in the Earth's
atmosphere and increased its content of primeval matter. Oro calculated
that as a result of this alone the Earth during the first 2 billion years of
its existence must have received from outer space 2.10^8 to 1×10^{12}
metric tons of comet material, basically consisting of carbon compounds. As a matter of fact, according to Oro's concept, the mass of the
comet matter captured by the Earth must still be significantly large,
since we can also add here the matter in the head of the comet escaping
direct collision but passing at a sufficiently near distance to the Earth
(80).

The fall of the so-called Tunguska meteorite occurring in 1908 is a
graphic example of collision with the Earth. As Fesenkov showed recent-

ly (81), this body, which remained a puzzle for a long time and which had fallen in the view of numerous witnesses and produced gigantic falls of trees in the taiga was, as a matter of fact, not a meteorite but the nucleus of a comet slowed down in the Earth's atmosphere at a height of 6–7 km, thus creating a powerful shock wave. The fact that the comet abounded not only in lightweight hydrocarbons but also in cyanogen is very significant (82). This simple compound of carbon with nitrogen must have entered into the composition of the original carbon compounds, which greatly enhanced the possibility of their further evolution.

However Miller and Urey (71) have calculated that the share of organic compounds borne by cometary matter is very small. According to their opinion, it is approximately 1/10,000 of that amount of the organics which were formed as the Earth's crust was formed. Neither could meteorites have given a sufficiently significant amount of carbon compounds. Thus apparently the basic mass of hydrocarbons and their close derivatives which were starting materials for the emergence of life were formed endogenously on the Earth's surface during formation of the Earth's crust. The further progress of evolution of these substances we can imagine, reproducing experimentally those conditions which existed at some time on the surface of the primordial Earth and studying how under these conditions the transformations of the original carbon compounds occurred.

REFERENCES

1. R. Russel and R. Farquhar. "Lean Isotopes in Geology." Wiley (Interscience), New York, 1960.
2. A. Vinogradova and A. Tugarinov. *Geokhimiya* 9, 723 (1961).
3. Y. Kulp. *21st Intern. Geol. Congr., Copenhagen, 1960, Rept. Session, Norden* Part 3, p. 18. Univ. of Copenhagen, Copenhagen, Denmark, 1960.
4. E. Anders and G. Goles. *J. Chem. Educ.* 38, 58 (1961).
5. E. Anders. *In* "The Moon, Meteorites and Comets" (G. Kuiper and B. Middlehurst, eds.), p. 402. Univ. of Chicago Press, Chicago, Illinois, 1963.
6. C. Patterson. *Natl. Acad. Sci. – Natl. Res. Council, Publ.* 400, 157 (1955).
7. F. Whipple. *Geochim. Cosmochim. Acta* 10, 230 (1956); *Proc. Natl. Acad. Sci. U.S.* 52, 568 (1964).
8. E. Rutherford. *Nature* 112, 305 (1923).
9. W. Fowler. *Proc. Natl. Acad. Sci. U.S.* 52, 524 (1964).

10. V. Baranov and K. Knorre. *Meteoritika* **21**, (1961).
11. E. Anders. *Rev. Mod. Phys.* (1961).
12. I. Reynolds. *Phys. Rev. Letters* **4**, 8 (1960).
13. I. Greenstein. *Proc. Natl. Acad. Sci. U.S.* **52**, 549 (1964).
14. W. Fowler. *Ann. Phys. (N.Y.)* **10**, No. 2 (1960).
15. F. Hoyle. *Observatory* (1962).
16. R. Brownlee and A. Cox. *Sky & Telescope* **21**, 251 (1961).
17. F. Hoyle, W. Fowler, G. Burbidge, and E. Burbidge. *Astrophys. J.* **139**, 909 (1964).
18. O. Struve. *9th Colloq. Intern. Astrophys., Liège, 1959* 5th Ser., Vol. 3. Cointe Sclessin. Inst. d'Astrophys., Soc. Roy. Sci., Liège, 1959.
19. A. Cameron. *9th Colloq. Intern. Astrophys., Liège, 1959* 5th Ser., Vol. 3. Cointe Sclessin. Inst. d'Astrophys., Soc. Roy. Sci., Liège, 1959.
20. J. Greenstein. *Sci. Progr. (New Haven)* [11] p. 173 (1963).
21. V. Fesenkov. *Priroda* No. 10, 2 (1964).
22. D. Frank-Kamenetskii. *Priroda*, No. 11, 17 (1963).
23. D. ter Haar and A. Cameron. *In* "Origin of the Solar System" (R. Jastrow and A. G. W. Cameron, eds.), p. 4. Academic Press, New York, 1963.
24. O. Shmidt. "Chetyre lektsii o teorii proiskhozhdeniya Zemli" ("Four Lectures on the Theory of the Origin of the Earth"), 3rd ed. Izd. Akad. Nauk SSSR, 1957.
25. E. Shauman. "Voprosy kosmologii" ("Problems of Cosmology"), Vol. 3, p. 227. Izd. Akad. Nauk USSR, 1954.
26. L. Goldberg, E. Muller, and L. Aller. *Astrophys. J.* Suppl. 5, No. 45, 1 (1960).
27. H. Brown. *Astrophys. J.* **111**, 641 (1950); H. Brown and M. Ingraham. *Phys. Rev.* **72** 347 (1947).
28. L. Goldberg, E. Muller, and L. Aller. *Astrophys. J.* Suppl. 5, No. 45, 1 (1960).
29. H. Suess. Summer Course on Nuclear Geology, p. 28. Varenna, 1960; G. Khil'mi. "200 let nauchnoi kosmogonii" ("200 Years of Scientific Cosmogony"). Izd. "Znanie," Moscow, 1955.
30. F. Hoyle. *Quart. J. Roy. Astron. Soc.* **1**, 28 (1960).
31. H. Urey. *In* "Space Science" (D. LeGalley, ed.), p. 4. Wiley, New York, 1963.
32. H. Urey. "The Planets, their Origin and Development." Yale Univ. Press, New Haven, Connecticut, 1952.
33. H. Urey. Hugo Muller Lecture. *Proc. Chem. Soc.* (1958).
34. V. Baranov. *In* "Vozniknovenie Zhizni po Vselennoi" ("Genesis of Life in the Universe"), p. 22. Izd. Akad. Nauk SSSR, 1963.
35. A. Vinogradov. *Geokhimiya* **1**, 3 (1961); **3**, 269 (1962).
36. P. Karlei. "Vozrast Zemli" ("Age of the Earth"). Fizmatizdat, 1962.
37. P. Harteck and I. Hensen. *Z. Naturforsch.* 3a, 591 (1948).
38. W. Latimer. *Science* **112**, 101 (1950).
39. V. Goldschmidt. *Skrifter Norske Videnskaps-Akad. Oslo, I: Mat-Naturv. Kl.* No. 4 (1937).
40. M. Rutten. "The Geological Aspects of the Origin of Life on Earth." Elsevier, Amsterdam, 1962.

41. K. Rankama. *Geol. Soc. Am., Spec. Papers* **62**, 651 (1955).
42. P. Ramdohr. *Abhandl. Deut. Akad. Wiss. Berlin, K1. Chem., Geol. Biol.* **3**, 35 (1958).
43. H. Lepp and S. Goldich. *Bull. Geol. Soc. Am.* **70**, 1637 (1959).
44. C. Sagan. *Radiation Res.* **15**, 174 (1961).
45. P. Merill. "Linii khimicheskikh elementov v astronomich. spekrakh" ("Lines of the Chemical Elements in Astronomical Spectra"). Carnegie Inst., Washington, D. C., 1956.
46. I. Plaskett. *Publ. Dominion Astrophys. Obs., Victoria, B.C.* **11**, No. 16, 72 (1924).
47. E. Waterman. *Lick Obs. Bull.* **8**, 1 (1913).
48. H. von Klüber. "Das Vorkommen der chemischen Elemente im Kosmos." Barth, Leipzig, 1931.
49. H. Kramers and D. ter Haar. *Bull. Astron. Inst. Neth.* **10**, 137 (1946).
50. D. Bates and L. Spitzer. *Atrophys. J.* **113**, 441 (1951).
51. F. Whipple. *Nature* **189**, 127 (1961).
52. B. Donn and H. Urey. *Astrophys. J.* **123**, 339 (1956).
53. A. Adel and V. M. Slipher, *Phys. Rev.* **46**, 902 (1934).
54. G. Kuiper. *Astrophys. J.* **100**, 378 (1944).
55. E. E. Krinov. "Meteorites." Pergamon Press, Oxford, 1960; "Principles of Meteoritics." Pergamon Press, Oxford, 1960.
56. H. Urey. *Nature* **192**, 1119 (1962).
57. L. Kvasha. *In* "Vozniknovenie Zhiznii vo Vselennoi" ("Origin of Life in the Universe"), Izd. Akad. Nauk SSSR, 1963; B. Masson. J. *Geophys. Res.* **65**, 2969 (1960).
58. G. Vdovykin. *Geokhimiya* No. 2, 135 (1962).
59. S. Vaughn and M. Calvin. North-Holland Publ., Amsterdam, 1960; M. Calvin. "Chemical Evolution." Coudon Lectures, 1961.
60. S. Vallentyne. Cited by S. W. Fox. *In* "The Origin of Prebiological Systems" (S. W. Fox, ed.), Academic Press, New York, 1965.
61. G. Mueller. *In* "Advances in Organic Geochemistry" (E. Ingerson, ed.), Monograph No. 15, Earth Sci. Ser., p. 119. Pergamon Press, Oxford, 1964.
62. M. Briggs. *Life Sci.* **1**, 63 (1963).
63. F. Aston. *Nature* **114**, 786 (1924).
64. H. Suess. *J. Geol.* **57**, 600 (1949).
65. H. Brown. *In* "The Atmosphere of the Earth and Planets" (G. Kuiper, ed.), p. 258. Univ. of Chicago Press, Chicago, Illinois, 1949.
66. H. Urey. *Astrophys. J.* Suppl. 1, 147 (1954).
67. J. Lederberg and D. Dowier. *Science* **127**, 1473 (1958).
68. W. Fowler, G. Greenstein, and F. Hoyle. *Am. J. Phys.* **29**, 393 (1961); *Geophys. J. Roy. Astron. Soc.* **6**, 148 (1962).
69. I. Glasel. *Proc. Natl. Acad. Sci. U.S.* **47**, 174 (1961).
70. J. Bernal. "Gorizont y biokhimii" ("Horizons of Biochemistry"). Izd. "Mir," 1964. J. Bernal. *In* "The Origin of Prebiological Systems." (S. W. Fox, ed.). Academic Press, New York, 1965.

71. S. Miller and H. Urey. *In* "Problemy evolyuts. i tekhnich. biokhimii" ("Problems of Evolution and Technical Biochemistry") (V. Kretovich, ed.), p. 357, 1964.
72. P. Kropotkin. *In* "Vozniknovenie Zhizni na Zemle" ("Origin of Life on Earth") (A. I. Oparin, ed.), p. 84. Pergamon Press, Oxford, 1959.
73. V. Sokolov. *In* "Vozniknovenie Zhizni na Zemle" ("Origin of Life on Earth") (A.I. Oparin, ed.), p. 84. Pergamon Press, Oxford, 1959.
74. I. Petersil'e. *Geokhimiya* 1, 15 (1962).
75. P. Kropotkin. See "Sovetskaya Geologiya" ("Soviet Geology"), Coll. 47, p. 104. Gos. Nauchno-tekhn. Izd., 1955.
76. R. Robinson. *In* "Advances in Organic Geochemistry" (E. Ingerson, ed.), Monograph No. 15, Earth Sci. Ser., p. 7. Pergamon Press, Oxford, 1964.
77. J. Bernal. *Proc. 5th Intern. Biochem. Congr., Moscow, 1961* Vol. III, p. 3. Pergamon Press, Oxford, 1963.
78. J. Oro. *Nature* 190, 389 (1961).
79. H. Urey. *Nature* 173, 556 (1957).
80. E. Opik. *Irish Astron. J.* 4, 84 (1956).
81. V. Fesenkov. *Astron. Zh.* No. 5 (1961).
82. P. Swings and L. Haser. "Atlas of Representative Cometary Spectra." Univ. of Liège, Astrophys. Inst., Louvain.

CHAPTER 3

FORMATION OF THE "PRIMITIVE SOUP"

How can we determine the pathway of chemical evolution of those carbon compounds which were formed endogenously on the Earth's surface or were carried here by meteorites or comets long before the origin of life?

At first glance it would seem that the most direct and satisfactory answer to this question must be given by geological and geochemical investigations – the direct discovery and study under natural conditions of those transformations of carbon compounds, which even now perhaps, are being carried out abiogenically somewhere without any direct relationship to living organisms. And even now gaseous hydrocarbons formed inorganically when the Earth's crust was formed pass into our atmosphere; meteorites and comets continue to "fatten up" the Earth with carbon compounds, just as in the remote past; waters of the modern ocean also contain small quantities of various organic materials. True, these substances originated biogenically from the decay of marine organisms, but perhaps somewhere in the depths of the ocean, in the absence of living matter, they once again undergo an evolution like that which occurred on the primeval nonliving Earth; perhaps we can find out about this evolution, studying it directly under natural conditions.*

However, all that we know about this question forces us to expect disappointment in our search for such a direct frontal attack on the problem, and although geochemical investigation would have a certain interest for us, the results never could be applicable to phenomena occurring in the pre-actualistic epoch, could never be used directly for a discussion of the processes which took place on the Earth's surface during the initial phase of its existence.

*The notion of a repetitive and recurring origin of life has been recently expressed by Keosian (1) in his book "The Origin of Life."

There is no doubt that the abiogenic transformations of carbon compounds on the Earth's surface and in its atmosphere and hydrosphere preceding the appearance of life were accomplished under conditions differing widely from those at present. Basically these differences are reduced to the following closely connected circumstances:

1. The absence on the primeval Earth of a radiation belt, which, according to the data of Uffen (2), was formed as a result of motion in the Earth's crust approximately 2.5 billion years ago. Prior to this, corpuscular radiation of the sun could freely reach the Earth's surface.

2. The absence in the pre-actualistic atmosphere of free oxygen, which excluded direct extensive oxidation of reduced carbon compounds.

3. Abundance of short-wave ultraviolet radiation permeating the whole atmosphere and reaching the Earth's surface. This created enormously greater opportunities for abiogenic photochemical processes, than can occur presently at the single but longer wavelength radiation reaching the Earth's surface.

4. The absence of living organisms and their complete metabolic systems which quickly draw into the orbit of their activity a variety of organic compounds.

At present, we do not have such conditions on the Earth's surface in a natural environment. The ionizing particles emitted by the sun for the most part are captured by the Earth's magnetic field and deflected, forming a belt of radiation. The entire terrestrial sphere is surrounded at a height of 30 km above its surface with an ozone shield, which obstructs the passage to us of short-wave ultraviolet radiation. The atmosphere, the upper layers of the soil, and the whole hydrosphere are very rich in free oxygen. As recent investigations of the Soviet expeditionary ship "Vityaz" (3) have shown, even water in the deepest depressions of the Pacific Ocean is saturated with oxygen to a considerable degree. We can find reducing conditions only in very exceptional cases, for example, in the Norwegian fiords.

But even here, as everywhere else, these waters are rich in organisms, especially the anaerobic microbes, and in the presence of organisms we cannot study abiogenic processes. In this respect organisms "mix up all the cards for us." They give off into the external inorganic medium specific biogenic substances which can be formed only in the process of highly organized metabolism, and on the other hand, they absorb or consume other substances involving them in a metabolism, which was lacking, of course, in the primeval abiogenic medium.

As we now recognize, Charles Darwin (4) wrote about this in one of his letters. "It is often said that all the conditions for the initial emergence of a living organism are present at any given time. But if (and oh what a big if) we could conceive in some warm little pond, with all sorts of ammonia, phosphoric salts, light, heat, and electricity, etc. present, that a protein compound was chemically ready to undergo still more complex changes; at the present day such matter would be consumed or absorbed, which would not have been the case before living creatures were formed."

Thus, although at first glance it seems paradoxical, we need to recognize that the main reason why life cannot be generated *now* under natural conditions is the fact that it has already been generated, and because of this, radical changes in conditions have occurred on the Earth's surface, making impossible any sort of long-term evolution of organic compounds by the methods by which this evolution was refined during the pre-actualistic epoch of the Earth's existence.

We are therefore required to form our opinions of this primary evolution mainly from laboratory experiments in which we attempt to create artificially those physical and chemical conditions which prevailed at some time on the Earth's surface.

For this purpose we start with the generally accepted conviction that the chemical reactivity of methane or other organic substance is the same in the past as it is now, the same in the primary or secondary terrestrial atmosphere as in the modern flask of the chemist, although this chemical capability will be achieved in different ways depending on conditions of the surrounding medium. Thus as we reproduce consciously and experimentally conditions in the pre-actualistic epoch, we should expect results from which we can interpret events in the remote past. Of course, at the same time, we must also try to utilize other possibilities in order to compare these conclusions with the facts which can be obtained by a study of natural materials and phenomena, not only on the Earth's surface but also on meteorites or other heavenly bodies.

The voluminous data accumulated up to the present through organic chemistry demonstrate the exceptionally high reactive ability of hydrocarbons and their nearest derivatives (5). Beginning with Schorlemmer (6), all contemporary organic chemistry has been regarded as the chemistry of hydrocarbons and their derivatives, because these compounds conceal in themselves astonishing and unique chemical possibilities. There is no doubt whatsoever that in the course of the many hundreds of millions of years which separated the time of formation of the Earth from the moment when life emerged on it, that the hydrocarbons on the Earth's surface fully utilized these possibilities.

In the conditions prevailing in the pre-actualistic period they could react either with each other or with water vapor, ammonia, hydrogen sulfide, and other gases of the secondary reducing atmosphere of the Earth, forming that diversity of complex compounds which are completely foreign to inorganic nature. These reactions of hydrocarbons and their derivatives could have been carried out in the pre-actualistic atmosphere and hydrosphere spontaneously at the expense of the inherent potential energy of the reduced carbon compounds; the rate of such a series of reactions would depend strongly on the presence of some sort of inorganic catalyst.

A good illustration of this is furnished by the recently published detailed paper of Bird (7). Numerous examples of organic syntheses catalyzed by metals were presented, which, in the author's opinion, could also occur on the primeval Earth.

However, external sources of energy must also have played an important part in the organic chemical reactions of this period; first among these was short-wave ultraviolet light, then radioactive emissions from the sun and the Earth's interior, spark and silent discharges in the atmosphere and finally the high temperature of volcanic eruptions.

Miller and Urey (8) give the following data (Table 4) for those sources of energy available on the primitive terrestrial surface which could have been used for primordial abiogenic syntheses.

TABLE 4
Energy Sources for Primary Abiogenic Syntheses

	$Cal/cm^2/year$
Ultraviolet light from the sun:	
2500 Å	570
2000 Å	85
1500 Å	3.5
Electrical discharges	4.0 V
Ionizing radiation	0.8
Heat from volcanic eruptions	0.13

The quantitative relationships of various sources of energy are very prominently displayed in this table, but the energy of cosmic rays and charged particles emitted by the sun is not sufficiently emphasized. At the present time most of the ionized particles of this type are retarded by the Earth's magnetic field. But before the formation of the radiation belt in the pre-actualistic epoch, the magnitude of this energy source was much larger than it is now, and thus ionizing radiation also must have played a significant part in the prebiological syntheses of organic matter on the Earth.

The use of these energy sources under artificial laboratory conditions simulating the primordial terrestrial surface has made it possible to synthesize diverse, sometimes very complex biologically important compounds in numerous experiments. Starting materials for these syntheses are such primitive compounds as methane, ammonia, and water vapor, as well as some of their closely related derivatives, for example CO, HCN, formaldehyde, acetaldehyde, thiourea, ammonium thiocyanate, etc. The papers published on these subjects have already reached a very high total. Organic acids and aldehydes, amino acids and polypeptides, amines and amides, sugars, specifically ribose and deoxyribose, purine and pyrimidine bases, nucleosides and nucleoside phosphates, polynucleotides, porphyrinlike substances and other biologically important compounds have been obtained as a result of this research. Many of these papers were presented at the Moscow Symposium on the Origin of Life on Earth in 1957 (9) and at the Florida Conference on the Origin of Prebiological Systems in 1963 (10).

They have been summarized in the detailed paper by Pavlovskaya and Pasynskii (11) as well as in a recently published review paper by the same authors (12). We can therefore limit our statements here only to the most important investigations, principally of protein and nuclein components.

Miller (13) in his well-known work published in 1953, obtained fundamental data on the formation of amino acids when a gas mixture, simulating the possible composition of the Earth's primary atmosphere, was acted upon by electrical discharges. Miller passed both spark and silent discharges (Fig. 8) into a constantly circulating mixture of CH_4, NH_3 H_2 and water vapor for a week, and found in the mixture: glycine, alanine, α-aminobutyric and α-aminoisobutyric acids, β-alanine, aspartic

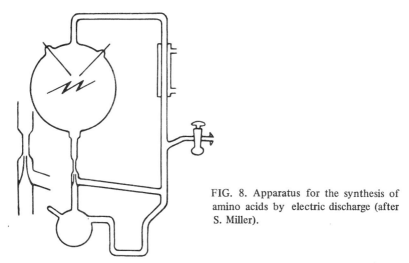

FIG. 8. Apparatus for the synthesis of amino acids by electric discharge (after S. Miller).

and glutamic acids, sarcosine, and N-CN$_3$-alanine. Intermediate products of the reaction were aldehydes and HCN.

Miller's data were confirmed by the work of Pavlovskaya and Pasynskii (14) using another type of apparatus and substituting surplus hydrogen for CO. The reason for this substitution was based on thermodynamic calculations. In subsequent work these authors demonstrated experimentally the sterility of the irradiated solutions and theoretically proved their stand that any reaction leading to the formation of aldehydes and HCN must have contributed to the formation of α-amino acids in the hydrosphere of the primeval Earth.

Abelson (15), using a more complex starting mixture of gases including CO, CO$_2$, and N$_2$ in addition to CH$_4$, NH$_3$, N$_2$ and H$_2$O, obtained a mixture of amino acids similar to Miller's, and also showed that NH$_3$ could not be replaced with N$_2$, nor CH$_4$ with CO$_2$. Oro (16), starting with C$_2$- and C$_3$-hydrocarbons and concentrated NH$_4$OH added to the list of amino acids, synthesized by spark discharge leucine, isoleucine, and valine. Heyns, Walter, and Meyer (17) introduced hydrogen sulfide into the starting mixture of gases, but were unsuccessful in synthesizing sulfur-containing amino acids, and obtained only thioacetate and thiourea.

Grossenbacher (18), who passed a spark discharge for 100–200 hours into a specially constructed apparatus containing a mixture of NH_3, CH_4 and H_2 in aqueous solutions, obtained the following mixture of amino acids: 2 moles of aspartic acid, 4 threonine, 14 serine, 16 glycine, 14 alanine, 4 lysine, 2 leucine, 2 isoleucine, and 1 glutamic acid. In addition he found peptides consisting of glycine and alanine (5:1) as well as glycine and isoleucine. At the end of the experiment the products of amino acid polymerization were isolated from the solution in the form of small multimolecular spheres or drops. There are fewer investigations on the synthesis of amino acids by ionizing radiation in a mixture of primeval gases. The experiments of Dose and Rajewsky (19) can be cited here; they found acid and neutral amino acids in a gas mixture of CH_4, NH_3, H_2, CO_2, N_2 and H_2O subjected to the action of X-rays. Hasselstrom, Henry, and Murr (20) obtained glycine and aspartic acid by irradiation with β-rays from a linear accelerator (2 MeV), while Paschke, Chang, and Yaung (21) demonstrated the formation of glycine and alanine from solid ammonium carbonate by ^{60}CO γ-irradiation (5 X 10^8 r). Palm and Calvin (22) using fast electron beams (5 MeV at a dose of 10^{10} ergs) on a mixture of methane, ammonia, hydrogen, and water vapor found glycine and aspartic acid.

The formation of amino acids by simple heating of the starting solutions was first discovered by Fox, Johnson, and Vegotsky (23). Oro and his co-workers (24) showed that several amino acids (glycine, alanine, serine, aspartic acid, threonine) were obtained simply by heating aqueous mixtures of formaldehyde and hydroxylamine for 40–60 hours at 80–100°C or even at lower temperatures. Similar results were obtained by Lowe, Ress, and Markham (25) who heated a 1.5 M solution of NH_3 and HCN at 90° for 18 hours and found a considerable quantity of amino acids; some were in the form of peptides which released after hydrolysis of the reaction product, leucine, isoleucine, serine, threonine, and glutamic acid in addition to glycine, alanine, and aspartic acid.

More recently Harada and Fox (26) carried out thermal syntheses from primeval gases in laboratory experiments simulating volcanic conditions. These experiments were very significant, and although the heat of volcanic eruptions is in last place in the table given above of energy sources, such syntheses could have a certain importance for the abiogenic formation of amino acids on the primitive terrestrial surface, since this form of energy was not distributed uniformly over the whole surface but was local in character.

But, of course, the energy of short wavelength ultraviolet rays, easily reaching the Earth's surface because of the absence in the pre-actualistic epoch of an ozone shield, was of basic significance in the primary abiogenic synthesis of organic compounds. The synthesis of amino acids and other organic compounds in experimental imitation of these conditions is now particularly well represented in the scientific literature of the world.

Groth (27) discovered that when a mixture of methane, ammonia, and water vapor was acted upon by ultraviolet light with wavelengths of 1470 and 1295Å (the resonance lines of a xenon lamp), glycine and alanine were formed, and when methane was replaced by ethane, α-aminobutyric acid also was formed. Terenin (28), for similar purposes, used a hydrogen lamp emitting ultraviolet light with a wavelength below 1300 Å, thus approximating the short-wave radiation spectrum of the sun. Irradiating a mixture of CH_4, NH_3, and CO in contact with water in the liquid state, he found alanine among the reaction products. Later, in Terenin's laboratory, Dodonova and Sidorova (29), irradiating the same mixture with ultraviolet light in the region of 1450–1800 Å, demonstrated the formation of glycine, alanine, α-aminobutyric acid, valine and leucine (or norleucine). The addition to the mixture of H_2S or CS_2 did not lead, however, to the formation of sulfur-containing amino acids, as was also the case when electrical discharges were used as a source of energy.

Pavlovskaya and Pasynskii (14), when they irradiated aqueous solutions containing 2.5% formaldehyde and up to 1.5% ammonium chloride or nitrate for 20 hours using ultraviolet light from a PRK-2 lamp, demonstrated the formation of the following amino acids – serine, glycine, glutamic acid, alanine, valine, phenylalanine and the basic amino acids (leucine, ornithine, and arginine). In the presence of adsorbents (bentonite, kaolin, limonite, and optical quartz) the same amino acids were found, with the exception of the basic acids, and in addition to these, isoleucine (30). In subsequent work, threonine was shown to be present, and when formaldehyde was replaced with acetaldehyde, arginine and tyrosine were also formed. Similar results have also been obtained by many other authors by the irradiation of mixtures of primitive gases and their derivatives with short-wave ultraviolet light (31). We should note that whether electrical discharges or ultraviolet irradiation are used for synthesis of primitive mixtures, not only amino acids, but also organic acids and aldehydes, amines and amides, and especially urea is formed (32).

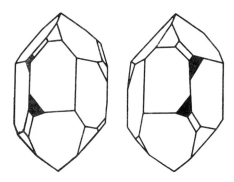

FIG. 9. Levo- and dextrorotatory crystals of quartz.

In addition to this, the monomeric molecules polymerize under these conditions, resulting in the formation of more complex compounds (33).

In connection with the abiogenic syntheses of amino acids and their polymers, it is necessary to touch briefly on a problem which has been discussed in the literature for a long time, that is, primary asymmetric synthesis. The formation of amino acids by the action of ultraviolet light opens up some theoretical possibilities.

Let us recall that the first practical asymmetric synthesis was carried out under laboratory conditions by a photochemical reaction caused by the action of circularly polarized ultraviolet light (34). It has also been shown that such light could have existed naturally on the Earth (35). Other possible ways are now known for producing asymmetric compounds outside of living nature. In particular, Bernal (36) has advanced the theory that asymmetry of organic matter could also come about abiogenically as a consequence of the synthesis of these compounds on the surface of nonsymmetrical quartz crystals (Fig. 9). This theory found its experimental verification in laboratory syntheses carried out by Terent'ev and Klabunovskii (37).

Wald (38) has expressed doubt, however, that any of these abiogenic factors could create conditions for the emergence of stable asymmetry. According to his opinion, the geochemical synthesis of organic molecules has produced only racemic mixtures from which individual optical isomers were selected during the formation of structures of a higher order – polypeptides and proteins from amino acids, nucleic acids from nucleotides, etc. In particular, according to Wald, this must have occurred when proteins and polypeptides forming α-spirals were selected.

Pasynskii (39) also imparts the greatest significance in the problem of the emergence of asymmetry to the appearance of conditions for the formation of stereospecific polymers with an ordered spatial arrangement of side groups which are reached by catalysts not having optical activity. It is possible, he points out, that optical asymmetry was developed not by way of the initial formation of optically active monomers and their transformation into optically active polymers, but, on the contrary, high molecular weight asymmetric catalysts were first formed as a result of stereospecific polymerization, and then, from this starting point, the optical asymmetry of low molecular weight compounds developed; this asymmetry thus had a secondary origin.

In addition to the researches on the abiogenic synthesis of amino acids and polypeptides, a great many papers have been recently published dealing with a similar synthesis of nucleotide components. Oro (40), subjecting a mixture of hydrogen cyanide, ammonia, and water to moderate heating (from 30° to 100°C) synthesized one of the purine bases, adenine. This experiment demonstrated that these complex heterocyclic compounds could be formed spontaneously from simple mixtures of primitive gases from the Earth's atmosphere, since cyanogen is easily formed from methane when it is irradiated in an aqueous ammoniacal system.

Oro also found among the reaction products in addition to adenine, 4-aminoimidazole-5-carboxamide (AICA), 4-aminoimidazole-5-carboxamidine (AICAI), formamide and formamidine. Apparently adenine is formed directly from AICIA when it condenses with formamidine (41). Another purine, guanine, was synthesized by Oro (42) from AICA and urea, which, as was shown above, is easily formed from primitive gases by ultraviolet radiation and electrical discharge. Among the pyrimidine bases Fox has synthesized uracil by heating malic acid and urea (43). Oro (44) obtained uracil by the reaction of urea in ammoniacal solution with unsaturated compounds formed from primitive gases by electrical discharges.

The carbohydrate components of the nucleotides – ribose and

2-deoxyribose – were synthesized under conditions similar to the primitive: first by Mariani and Torraca (45) and second by Oro (46) by condensing formaldehyde and acetaldehyde in the presence of basic catalysts.

Extensive opportunities for the abiogenic synthesis of nucleotide components also are found in the use of ionizing radiation, ultraviolet light, and electrical discharges.

According to Calvin (47), the radioactive emissions of several elements, ^{40}K in particular, (which was very abundant in the primitive crust of the Earth) must have been to a sufficient degree an important source of energy for abiogenic syntheses. Further, M. Calvin and his co-workers (48) imitated the radiation of ^{40}K acting on a primitive gas mixture with very high energy electrons. Ponnamperuna et al. (49) using this method obtained adenine from methane, ammonia, and water. Guanine, cytosine, uracil, and thymine are not formed under these conditions. It is interesting to compare this fact with the preeminent position occupied by adenine in all biological systems – which indicates its great antiquity in the history of the development of life. Molecular orbital calculations show that of all the biologically important purines and pyrimidines, adenine has the highest resonance energy (50) which, of course, makes its abiogenic synthesis the most probable.

Ribose and deoxyribose, and possibly other sugars also, are obtained from the same mixture of methane, ammonia, and water by the action of ionizing radiation. The synthesis of sugars has been particularly successful when formaldehyde is incorporated into the original mixture; formaldehyde is easily formed from the primitive gases by electrical discharges. Ribose and deoxyribose have been synthesized by Calvin's group (49) from a mixture of gases acted upon by ultraviolet light or the γ-radiation of ^{60}Co. When HCN was included in the original mixture, and when ultraviolet light was the active agent, adenine, guanine, and urea were synthesized (51).

A very important step in the further evolution of organic matter on the pathway to the origin of life was the abiogenic union of individual components into nucleotide complexes and the subsequent polymerization of these components into polynucleotides, just as amino acids had to be polymerized into polypeptides and proteinlike substances under the conditions existing on the primitive terrestrial surface.

The feasibility in principle of such a far-reaching abiogenic polymerization of organic substances has been confirmed by a number of model experiments in which primarily amino acids or compounds related to them were used as monomers.

We should mention to begin with, the research of Akabori (52), who, as early as 1955, postulated the abiogenic synthesis under conditions like those on the primitive Earth of proto-proteins (that is, amino acid polymers having a random arrangement of amino acid residues in the polypeptide chain). Soon thereafter he demonstrated this synthesis experimentally (53), obtaining from formaldehyde, ammonia and HCN aminoacetonitrile, the polycondensate of which was absorbed on kaolin with the formation of polyglycine. Side chains could then be introduced into the polyglycine by means of a reaction with aldehydes or with unsaturated hydrocarbons. Akabori successfully produced polyglycine and its analogs with a molecular weight of approximately 15,000 by these methods.

Wilson (54), conducting experiments similar to Miller's, passed a silent discharge into a specially constructed apparatus through an atmosphere of methane over an aqueous solution of yeast ash, ammonia, and hydrogen sulfide. A polymeric film was thereby formed; infrared analysis of the film indicated the presence of a polyethylene type of structure with C-O bonds. This material was hydrophilic, it absorbed methylene blue and fuchsin from aqueous solution. If ammonia and hydrogen sulfide were not used in the experiment hydrophobic films were formed.

The experiments of Fox and his co-workers are of fundamental interest (55): they used thermal polycondensation on a mixture of various amino acids. They showed that at a temperature on the order of $150°-180°C$ in the presence of considerable quantities of aspartic and glutamic acids, spontaneous copolymerization of 18 different amino acids occurs, leading to the formation of polypeptides with molecular weights of from 3000 to 9000. Fox called them proteinoids. According to color reactions, infrared spectra, solubility, electrophoretic mobility, as well as susceptibility to enzyme attack, Fox could not distinguish them from such proteins as casein, for example. They were isolated from a 1% aqueous solution of NaCl in the form of globules visible under the microscope − microspheres − to a more detailed examination of which we will turn later.

Phosphorus compounds have apparently played a special role in the processes of primordial polymerization. Phosphoric acid is completely condensed at temperatures of about 300° and higher and polyphosphate is formed. On the surface of the primitive Earth with its widely occurring local hot spots, polyphosphates and their organic compounds must have been formed on quite a large scale.

At the present time inorganic polyphosphates are found in considerable amounts in lower organisms: in bacteria, algae, fungi, some protozoa and insects. These are linear polymers of orthophosphoric acid with a molecular weight of 30,000–100,000. Cyclic trimeta- and tetrametaphosphates have also been isolated from several materials (56). Apparently, inorganic polyphosphates are significant in the metabolism of lower organisms in contrast to more highly organized living beings in which their presence is not characteristic.

Kulaev and Vagablov (57), on the basis of their investigations of the green alga *Scenedesmus obliquus* consider it proved that phosphorylation is entirely responsible for the syntheses of inorganic polyphosphates found in the organism; the phosphorylation occurs during an anaerobic fermentation of glucose similar to that in alcohol fermentation or glycolysis which results in the accumulation of energy in ATP. On the other hand, Szymona and Szymona (58) isolated an enzyme from lower organisms which without the action of ATP transfers high energy phosphate from polyphosphate to glucose with the formation of glucose 6-phosphate.

Belozerskii (59) at the Moscow Symposium in 1957 advanced the theory that inorganic polyphosphates could be significant both in abiogenic syntheses and in primordially forming life, to a certain degree replacing ATP.

Schramm and his co-workers (60), using esterified condensed phosphates in nonaqueous medium at a temperature of about 55–60°C carried out a polycondensation of various biologically important compounds. They obtained polypeptides from alanine glycinylglycine, polyarginine from arginine, polysaccharides with a molecular weight of 50,000, nucleosides, nucleotides and finally polynucleotides – polyadenine (molecular weight 21,000), polyuracil (molecular weight 50,000), etc.

On the basis of these experiments Schramm advanced the theory that similarly, through the agency of polyphosphates, proteinlike and nucleoproteinlike polymers could have formed on the primitive Earth's surface; these polymers, however, should be distinguished from modern proteins

and nucleic acids by the more or less random arrangement of monomeric residues in their chains.

Somewhat later, Schramm (61) defined more specifically the composition of esterified polyphosphates necessary for his syntheses. This was
a mixture of cyclic ethylmetaphosphates and linear ethylpolyphosphates;
for successful syntheses it was necessary to have 70% tetraethyltetra-
metaphosphate and 30% tetraethyltetrapolyphosphate.

However, at the conference on the origin of prebiological systems
held in Florida in 1963, Schwartz (62) showed that it was also possible
to synthesize polynucleotides by means of free polyphosphoric acid in a
nonaqueous medium at 60°C, that is, on a purely inorganic basis.

All this information leads us to speculate that at the dawn of the
origin of life on Earth, inorganic polyphosphates could readily have
participated in the formation of simple mechanisms joining the energy-
giving and energy-receiving reactions. However, it is necessary to keep in
mind that all these syntheses mentioned above were carried out in a
nonaqueous medium.

Ponnamperuna *et al.* (63) used ethylmetaphosphate for the synthesis
of nucleotide components in aqueous medium. Ultraviolet light with a
wavelength of 2400–2900 Å, which was readily available on the primitive terrestrial surface, served as an energy source. The syntheses were
conducted in aqueous solutions of adenine, adenosine, adenylic acid,
ribose, and ethylmetaphosphate. The mixtures were irradiated with ultraviolet light at a wavelength of 2573Å at 40°C. Under these conditions
adenine was transformed to adenosine, adenosine into AMP, AMP into
ADP, and finally ADP into ATP. Thus the feasibility was demonstrated
of an abiogenic synthesis under premodern terrestrial conditions of
adenosine triphosphoric acid, that basic "energetic currency" of living
organisms.

We can imagine that ethylmetaphosphate was not the only nor the
earliest source of phosphorus in primitive times. Adenosine can be
formed from adenine and ribose by the action of ultraviolet light not
only in the presence of ethylmetaphosphate but also of inorganic phosphoric acid. Ethylmetaphosphoric ester is necessary, however, for the
synthesis of adenosine phosphates under the same conditions.

The use of ionizing radiation for these purposes gives still more
promising results. With this energy source the synthesis of adenosine in
an aqueous solution of adenine and ribose can go even in the absence of

phosphorus, while the mono-, di-, and triphosphonucleotides require for their synthesis the presence of inorganic phosphate alone (64).

Summing up all these results, we see that model syntheses carried out under conditions simulating the chemical and energetic environment on the surface of the primitive Earth gives us quite convincing material for an opinion regarding the primary abiogenic reactions of the original carbon compounds in the pre-actualistic epoch of our planet's existence.

Much less significant are data obtained by direct observation of organic reactions under natural conditions, both ours on Earth and those on other cosmic bodies. Investigations into the organic reactions of petroleum would be very valuable to us in this respect, but here the abiogenic and biogenic processes are so closely interwoven that it is difficult to discriminate between them, although lately there have been some promising indications in this direction (65).

The same can be said with respect to the evolution of organic matter beyond the Earth's limits. Although the initial stages of this evolution, as we have seen above, are universal and can be found on the most diverse cosmic bodies, succeeding transformations of carbon compounds are much more specific in nature, closely related to the unique development of a given cosmic body. Our investigations of the more complex organic compounds, their formation and transformation, is limited to bodies of our solar system alone, and then only to that region where terrestrial planets and asteroids were formed.

The best factual data on the nature and reactions of the more complicated extraterrestrial organic matter we obtain principally by a study of the meteorities, particularly the carbonaceous chondrites (66). Here the synthesis of complex organic compounds must have occurred in a reducing medium at relatively cold temperatures and very low gravitation. Unlike the Earth, the carbonaceous chondrites contain water only in the bound state, in the form of hydrated rocks. Their organic matter consists mostly of relatively high molecular weight hydrocarbons.

Hydrocarbons were first recorded by Berzelius (67) in 1834 in the composition of the Alais meteorite. Somewhat later Wöhler (68) isolated from a chondrite falling on Hungary near Kaba a quantity of polymeric hydrocarbon resembling ozocerite. Later bituminous material was extracted from most of the carbonaceous chondrites now known (69).

In particular, Mueller (70) lately extracted from the Cold Bokkeveld meteorite, a polymeric organic substance which he characterized as bitumen. The optical neutrality of this material caused Mueller to conclude that it had an abiogenic origin. It is interesting that treatment of the original meteoritic material with boiling hydrofluoric acid, in which silicates usually dissolve readily, had no effect in this case. The reason for this abnormality Wilson (71) sees as the fact that the silicate particles are coated on the surface with a thin film of organic polymers similar to those which Wilson obtained in his experiments described earlier.

Mass spectroscopic analysis, ultraviolet and infrared data, chromatography and other modern methods have enabled us to define more precisely the organic matter in carbonaceous chondrites. The aliphatic character of the hydrocarbon polymers has been determined, their ozoceritelike characteristics, the presence of sulfur and oxidizing compounds have been confirmed (72). In the USSR the bitumens of the Mighei and Groznaya carbonaceous chondrites have been investigated. Highly polymerized aliphatic hydrocarbons with high molecular weight straight chains of the paraffin type have been found in them (73).

Nagi *et al.* (74) subjected to mass spectroscopic analysis a high temperature distillate of the Orgueil carbonaceous chondrite and found aliphatic hydrocarbons containing 15-24 carbon atoms in the chain. These authors even compare them with the fatty acids of natural fats. However, the abiogenic formation of this meteoritic organic chemical has now been confirmed beyond scarcely any doubt.

Thus we see that the investigation of meteorites confirms our idea of the wide-ranging complexity and abiogenic polymerization of organic matter even though it has occurred on meteorites under conditions somewhat different from those on Earth, and hence has led to the formation of relatively homogeneous mixtures of polymeric hydrocarbons. The evolution of organic matter on our natural satellite – the moon – must have occurred under extremely unusual conditions, due mainly to the absence of an atmosphere and hydrosphere. Sagan (75) on the basis of a series of theoretical considerations has stated the theory that there are abiogenic organic chemicals on the moon. According to Sagan, they could have been formed on the moon or have been brought to its surface by meteorites and then subjected to a different type of transformation by the action of ultraviolet light and ionizing radiation, which is free to reach the unprotected surface of the moon. Sagan assumes that the organic matter on the moon is deposited under a layer of rock which protects it from extensive photochemical breakdown, at a depth of approximately 10 meters. Here according to Sagan, lies a layer of hydrocarbon compounds at a density of 10 gm/cm^2.

Wilson (71) goes even farther and postulates that the so-called "seas" of the moon are asphalt lakes formed by the discharge of abiogenic petroleum on the surface of our satellite from which the volatile components have then been evaporated, and under the influence of the sun's photons, asphalt and petroleum coke have formed. Of course, all this is still only a more or less daring hypothesis.

The space probes, launched by the USSR and the USA, and soft landings at various places on the moon, have revealed the microstructure of the surface of our natural satellite. Its soil has a homogeneous crumbly structure – it is granular matter in which the size of the grains is highly variable, but their consistency is reminiscent of wet sand on the Earth's surface (71a). However, we still do not have reliable information regarding the presence of organic matter on the lunar surface or in layers lying beneath the surface. We will have this information only when we analyze samples of materials obtained directly from the moon.

Recently the hope of learning more about the possibility of extraterrestrial life has rested on the study of our closest neighbor, Venus. This is because Venus is very favorably located with respect to the sun; it was formed in the same region of the proto-planetary cloud as the Earth, and has almost an exact circular orbit and a mass close to the mass of the Earth (0.83). All this, it would seem, must have brought about the same course of evolution of carbon compounds as on the Earth. However, at the present time we are apparently obliged to renounce this hope.

Venus has quite an extensive atmosphere, but its surface is hidden from our direct telescopic observation by a dense layer of high clouds. Thus conditions on its true surface were for a long time impossible to ascertain and only the use of radio astronomical methods and the launching of cosmic rockets (76) has shed some light on this question.

Data obtained from the Soviet automatic station "Venera 4" which reached that planet in October 1967, is of great significance in this respect.

First of all, the absence of an appreciable magnetic field and radiation belts on the planet have been established by the flight of the Venus station. Entering its atmosphere, the station lowered by parachute a special scientific laboratory. Measurements made by this laboratory indicated that close to the planet's surface a temperature of about 280°C was the rule on Venus, and the atmospheric pressure was 15 times that of the Earth. The atmosphere consists almost entirely of carbon dioxide gas; oxygen and water vapor together make up only one half of 1% of its chemical composition. Appreciable traces of nitrogen were not found.

In the light of all these data, it is very difficult to conceive of life forming on Venus.

Only completely oxidized carbon was found in its atmosphere. The formation and subsequent conversion of reduced carbon compounds which led to the emergence of life on our planet, apparently did not take place on Venus. Consequently the evolution of carbon compounds in this case must have followed completely different pathways than on the Earth because of the very different thermal histories of these planets.

Fesenkov (77) earlier postulated that Venus from the very beginning did not attain an appreciable rate of revolution and formed without gravitational heating. Being an initially comparatively cold body, Venus was not subjected to internal stratification. Its long-lived radioactive elements remained distributed more or less equally over the whole mass and were not carried up to the surface layers as a consequence of their affinity with the light silicates as was the case on the Earth. Because of this there was no general or local surface heating; neither a crust formed nor a secondary reducing atmosphere abundant in water vapor, methane, ammonia, etc.

Radioactive heat could cause only slow but constant warming of the whole planetary mass, which has continued up to the present and which accounts for its high temperature. Thus, a magnetic field could not form around Venus, oceanic basins could not arise and the carbon dioxide being formed in its interior accumulated in the atmosphere since it could not be absorbed by silicates and form carbonates.

More promising to us with regard to the question which interests us is the study of Mars (78), although its mass is considerably less than the mass of the Earth (0.11 compared to the Earth) and its surface temperature, because of its great distance from the Sun, is considerably lower than on our planet. It experiences a considerable daily fluctuation, varying in the tropic region from 20°C at noon to −65°C at night. The Mars day is similar to the Earth's, but the year is considerably longer

(687 days). The inclination of the equator to the plane of its orbit is similar to that of the Earth; a seasonal change analogous to ours occurs therefore.

The atmosphere of Mars is very thin. It consists for the most part apparently of nitrogen (98%), argon (1.2%), CO_2 (0.25%), and a small portion of water vapor. Feathery clouds of ice crystals and yellow clouds of dust or fine sand can be found in the atmosphere of Mars. White caps consisting of a very thin layer of ice cover the poles of Mars. In the spring along the edge of the polar caps and extending in the direction of the equator, dark, greenish-blue contiguous regions are formed, which gradually widen with the movement of the spring thaw of water. These regions are treated by many astronomers as zones of vegetation. However, the problem of the existence of life on Mars cannot be considered as resolved at the present time.

The low average temperatures, the small quantity of water, the high incidence of ultraviolet radiation on the Martian surface, of course are serious obstacles to the existence on Mars of the overwhelming majority of terrestrial organisms. However, many authors nowadays consider that Mars had a more abundant water supply in the initial stages of its existence than now; the evolution of organic matter and the emergence of life could have occurred here as it did on the Earth. In addition to this, the great ability of organisms to adapt to ambient conditions make it probable that life, once having arisen on Mars, might have been able to adjust to the extremely bleak climate which now prevails there.

From this point of view, the fact discovered by Sinton (79) is very interesting: in the infrared region the light from Mars has three bands at 3.45, 3.58, and 3.69μ which are characteristic of C-H bonds and indicate the presence of organic molecules, hydrocarbons in particular. These bands are peculiar only to the dark regions of the planet and are lacking in its light areas. Colthup (80) interprets these bands as proof of the presence on Mars of organic aldehydes, specifically acetaldehyde, which could have been formed there by the metabolism of anaerobic organisms just as it is in certain stages of alcohol fermentation. Sinton also defends the biogenic nature of organic matter on Mars since he considers that if constant regeneration of these compounds did not occur by metabolism, they would very rapidly be covered up by dust or destroyed by ultraviolet light.

In contrast to this, Rea (81) on the basis of his investigations, considers that the light bands of Sinton still cannot be regarded as proof of the presence of life on Mars and that it is necessary to consider preferable the abiogenic concept of organic matter on this planet.

Not so long ago, Young, Ponnamperuna, and McCaw (82) performed a model experiment demonstrating how organic substances and their polymers could have formed abiogenically on Mars. The authors constructed a chamber in which conditions analogous to the Martian were maintained: the temperature shifting in the sidereal cycle from $-70°$ to $+30°$; pressure equal to 65 microbars, an atmosphere consisting of 65% CO_2, 33% N_2, (supplemented when necessary by 2% acetaldehyde), soil consisting of limonite and sand, and a small quantity of water vapor. The chamber was illuminated by a quartz ultraviolet lamp (2537Å), with the illumination, as well as the temperature, changing in a daily rhythm.

The experiments showed a considerable fixation of CO_2. With the formation of organic substances, particularly the tentative formation of acetaldehyde. The acetaldehyde itself under these conditions polymerized easily, being changed into nonvolatile compounds, specifically sugars (pentoses and hexoses). The presence of limonite was highly significant; it adsorbed the products of synthesis and prevented their reverse breakdown by ultraviolet light.

According to the opinion of these authors, the organic compounds synthesized abiogenically in the Martian atmosphere (among them acetaldehyde) are built up on the planet's surface, accumulating in its dark regions, while in the light regions they can only be destroyed, which would explain the phenomenon of Sinton. It can be hoped that all the growing successes in understanding the cosmos will permit us soon to obtain very significant reliable data on the evolutionary pathways of organic compounds on extraterrestrial bodies of our solar system.

The specific characteristic of the conditions under which the abiogenic synthesis of the whole range of complex organic matter occurred on the Earth's surface was the presence there of a vast hydrosphere, completely lacking on the asteroids, moon, Venus, and Mars. Any aqueous expanse adjacent to the atmosphere must have created very favorable conditions for the single direction of the evolutionary course of organic matter.

It is necessary to keep in view that in all these syntheses of complex organic compounds from primitive gases, which have been simulated in the model experiments described above, the rapid removal of the final reaction products from the reaction zone under natural conditions was extremely important for a quantitative yield and the avoidance of destruction of the compounds formed.

Hull (83), for example, considers that ultraviolet light, being the most powerful source of energy, could not only create organic compounds but also break them down, and that equilibrium in a closed system will not be directed in favor of synthesis. Thus, according to his calculation, the concentration of organic compounds on the surface of terrestrial bodies of water can only be minutely small.

However, these calculations hold true only if the whole process is limited to a thin surface layer of water and the synthesized compound lacks the ability to move away from the reaction zone. Under natural conditions we do not have such an ideally closed system and the products of synthesis without fail move away from the influence of short-wave radiation. Very favorable conditions are obtained in this respect where various reaction zones are in contact, for example, the water surface adsorbing the products of ultraviolet radiation and the atmosphere; the interface between the cold atmosphere and the electrical discharge zone; the water currents and cold gases, rapidly carrying away substances formed on the hot surface of volcanic lava, etc. The accumulation of compounds formed was the most significant in the seas and oceans, into which these compounds passed from the atmosphere and surface layers of the lithosphere, and were stored at sufficient depth to prevent their reverse destruction by short-wave radiation. For these reasons, the quantity of organic compounds synthesized in the atmosphere and deposited in the waters of the terrestrial hydrosphere must have been quite considerable. According to the calculations of Urey (84) and Sagan (85) these deposits amounted to approximately 1 kg/cm^3 of surface, during the course of a billion years and their concentration in the oceans must have reached something on the order of 1%.

Thus, at a certain period of the Earth's existence, these waters were converted into a unique "primitive soup" containing, in addition to inorganic salts, a variety of organic substances — simple and complex monomers and polymers, particularly the energy rich phosphoorganic compounds, capable of many interreactions.

The composition of this "primitive soup" was changing all the time, undergoing evolution both as a whole and in individual parts. The organic matter it contained on the one hand, was constantly being replenished at the expense of endogenic and exogenic sources of carbon compounds (the Earth's crust, meteorites and comets) and on the other hand, was being diminished as a result of partial but extensive breakdown. In this connection, the concentration of water-soluble organic substances varied widely not only in its entirety but particularly in individual more or less isolated waters where it could, for example, increase by local evaporation of water. Besides these quantitative changes, the qualitative composition of the "primitive soup" also underwent evolution, its contents becoming more complex and polymerizing; new compounds also arose, particularly as conditions on the Earth's surface changed.

The recent experiments of Szutka (86) on the abiogenic synthesis of porphyrins may serve to illustrate this. On the basis of these experiments Szutka concludes that porphyrins and their derivatives must have been generated at a relatively late stage in evolution at the beginning of the transitional period from the pre-actualistic to the actualistic epoch, when a small amount of free oxygen had appeared in the Earth's atmosphere and the ozone shield had formed (87).

It is interesting to compare this conclusion with investigations showing that porphyrin derivatives do not participate in the metabolism of a number of present-day anaerobes (88), while they are a very important component of the oxidative enzymes of aerobes.

Whatever course the evolution of the "primitive soup" pursued, in principle, during the whole period, it remained a complex solution of inorganic salts and organic matter. The order of organic chemical reactions taking place in it, the formation and breakdown of its organic chemicals differed fundamentally from the order which is characteristic for all living matter without exception. In living things, due to a certain preexisting organization, the succession of individual reactions is strictly brought into conformity with a single metabolic process (89). The order of the processes is extemely purposeful and is capable of leading to the constantly repeated synthesis of extremely complex and specific compounds which in this way can be generated and accumulated in the cell in significant quantities.

In the "primitive soup" such an orderly progression was of course entirely lacking. As is the case in a simple aqueous solution of organic substances, the chemical transformations in the "soup" took place in accordance with only the general laws of chemistry and physics. They were not directed or organized and ranged over the widest fields of chemical possibilities. Any substance could be altered here in the most diverse ways, and individual reactions crossed and recrossed each other in a completely whimsical fashion. In this way a large variety of all possible kinds of organic compounds and their polymers could arise, but the more complex and specific a given substance was, the greater the number of successive reactions which had to participate in its formation, the less probable would be its formation, and consequently, the less the concentration of this specific substance in the "primitive soup."

Thus the wide-ranging abiogenic formation of sugars, amino acids, purine, and pyrimidine bases and their nonspecific polymers was easily possible, but the abiogenic formation of modern proteins and nucleic acids was extremely improbable; that is, the formation of substances having a molecular structure with a specific, strictly determined arrangement of amino acid or mononucleotide residues, a molecular structure perfectly adapted to those functions which proteins or nucleic acids carry out in any living body. Proteins, particularly, have such a function in their specific catalytic activity as enzymes in the living cell.

As we have seen above, we are still far from establishing an abiogenic synthesis for all amino acids composing modern proteins. We must assume, then, that some amino acids were formed later under more complex conditions during period of higher evolutionary development of organic matter. If this is so, then the primitive polypeptides and proto-proteins must not have contained a complete selection of amino acids. They were much more primitive polymers than modern proteins even in their amino acid composition. Yet this alone could limit their catalytic function, since such amino acids as histidine or cysteine, for example, are obligatory components of the active centers for many modern enzymes.

But the most significant feature of the enzymic activity of modern proteins is the strictly regulated arrangement of amino acid residues in their polypeptide chains and the particular way of packing this chain into a protein globule.

Such an ordered intramolecular structure is formed now only as a result of the action of very complex, highly organized chemical mechanisms of living cells and the formation of these mechanisms was absolutely impossible in the organic matter simply in solution in the "primitive soup." Proteins similar to modern highly organized ones could not have been formed there. This is particularly true of enzymes (90) which are unique organs of the living cell on the molecular level. Their intramolecular structure is excellently, "appropriately" adapted to the fulfillment of those catalytic functions which they perform in the metabolism of the intact living system. But it is the appearance of this adaptation in a simple aqueous system, the "primitive soup," prior to the formation of such complete systems, which is extremely improbable.

This concept of the initial formation of enzymes is reminiscent to a certain degree of the concept of the ancient Greek philosopher Empedocles (91) according to which individual organs were initially formed on the Earth — "out of it many heads without necks appeared, arms wandered unattached to shoulders, eyes without faces strayed about alone." Later these organs became attached to each other, as it were, and thus animals and humans were formed. Now we know that eyes or arms can arise only during the evolution of an intact organism. In a similar manner, a properly constructed protein enzyme could be formed and perfected only in the process of evolution of an entire system already having some form of metabolism, even though it was primitive, in which enzymes played their specific role. In the process of evolution, it was in this system alone, having a form of metabolism, in which a stepwise adaptation of the internal structure occurred, at first of the primitive proteins and then of the enzymes proper, to those functions which they perform in metabolism.

Consequently, in order to solve the problem of the initial genesis of an ordered protein structure suited to its purpose, it is first of all necessary to determine how a definite sequence of metabolic reactions could have been formed from the chaos of criss-crossing reactions; it is necessary to visualize clearly and to prove experimentally methods of formation in the "primitive soup" of such initial systems in which, by an evolutionary process, a definite sequence of interactions with the external medium could be built up, gradually approaching the modern characteristics of all living metabolism.

At the present time in the scientific literature, the opinion has been widely expressed that these initial systems could be simply individual molecules of polynucleotides with a random arrangement of monomeric residues in the chain which originally emerged from the universal "soup." They have inherent in them the complementarity characteristic of all polynucleotides. Thus, according to Haldane (92), Schramm (61), and other authors (63), under abiogenic conditions there must have arisen the tendency toward a gradual, more rapid synthesis of preexisting polynucleotides, toward an accelerating "self-multiplication" of molecules having a definite secondary structure of the polymeric chain. Subjected to mutations and natural selections, the polynucleotide molecules continuously evolved and perfected their secondary structure, approaching in this way the structure of modern nucleic acids.

However, on the basis of such evolution "at the molecular level" it is difficult to visualize, much less experimentally reproduce, the origin of metabolism. If we produce this type of mononucleotide polymerization in pure isolated solutions, this leads only to the formation of peculiar aggregates, resembling polynucleotides in their structure, which under natural conditions would have only formed deposits similar to deposits of ozocerite or other mixtures of organic homologs. If this polymerization were accomplished in the presence of other polymers, polypeptides, for example (which, apparently, must have taken place in the "primitive soup"), the resulting polynucleotides inevitably would form with these polymers multimolecular complexes isolated from the surrounding solution in the form of individual systems (coacervate drops), as we have recently shown in experiments in our laboratory at the A. N. Bakh Institute of Biochemistry of the USSR Academy of Sciences (94). It is only the magnitude of the polymeric molecules formed which is essentially significant in this type of isolation and not the sequence of the monomeric residues in their chains; the arrangement can be completely random.

It is important to note that in connection with the formation of complex systems the equilibrium of the polymerization reactions is abruptly shifted in the direction of synthesis.

Precisely these complex systems, and not individual molecules in solution, must have been the basis of the original formations which later became the first organisms. Only this sort of system could have undergone natural selection evolution based on its interaction with the external medium. Initially these systems included only primitive randomly structured polypeptides and polynucleotides. But later on these latter came to acquire a more and more well-ordered intermolecular structure, adapted to those functions which made them a part of complete systems undergoing evolution. Outside of these systems it is impossible to perfect individual molecules, just as it is impossible to form and perfect any sort of organ — eye or arm — outside the organism.

The emergence of multimolecular complex systems in the initially homogenous "soup" of the terrestrial hydrosphere and their subsequent development is a further very important stage in the evolution of carbon compounds on the route to the origin of life.

REFERENCES

1. J. Keosian. "The Origin of Life." Reinhold, New York, 1964.
2. R. Uffen. *Nature* **198**, 143 (1963).
3. V. Bogorov. "Dal'nie plavaniya na Vityaze." Izd. "Znaniya," 1961.
4. "The Life and Letters of Charles Darwin" (F. Darwin, ed.), Vol. 3, p. 18. Murray, London, 1887.
5. A. Oparin. "Vozniknovenie Zhizni na Zemle" ("Origin of Life on Earth"), 3rd ed., Chapter 5. Academic Press, New York, 1957.
6. C. Schorlemmer. "The Rise and Development of Organic Chemistry," Rev. ed. London, 1894.
7. C. Bird. *Chem. Rev.* **62**, 283 (1962).
8. S. Miller and H. Urey. *Science* **130**, 245 (1959).
9. Proc. Intern. Symp. Origin of Life on Earth, Moscow, 1957. Pergamon Press, Oxford, 1959.
10. *Proc. Conf. Origin Prebiol. Systems, Wakulla Springs, Florida,* 1963. Academic Press, New York, 1965.
11. T. Pavlovskaya and A. Pasynskii. *In* "Problemy evolyutsionnoi i tekhicheskkoi biokhimmi" ("Problems of Evolutionary and Technical Biochemistry"), p. 70. Izd. "Nauka," 1964.
12. A. Pasynskii and T. Pavlovskaya. *Usp. Khim.* **33**, 1198 (1964).
13. S. Miller. *Science* **117**, 528 (1953); *J. Am. Chem. Soc.* **77**, 2351 (1955). *In* "Origin of Life on Earth" (A. I. Oparin, ed.), Pergamon Press, Oxford, 1959; *Biochim. Biophys. Acta* **23**, 480 (1957).
14. T. Pavlovskaya and A. Pasynskii. *In* "Origin of Life on Earth" (A. I. Oparin, ed.). Pergamon Press, Oxford, 1959.
15. P. Abelson. *Science* **124**, 935 (1956); *Carnegie Inst. Wash. Year Book* **55**, 171, (1956); *Ann. N.Y. Acad. Sci.* **69**, 274 (1957).
16. J. Oro. *Nature* **197**, 862 (1963).
17. K. Heyns, W. Walter, and E. Meyer. *Naturwissenschaffen* **44**, 385 (1957).
18. K. Grossenbacher and C. Knight. *In* "The Origin of Prebiological Systems" (S. W. Fox, ed.), p. 269. Academic Press, New York, 1965.
19. K. Dose and B. Rajewsky. *Biochim. Biophys. Acta* **25**, 225 (1957).
20. T. Hasselstrom, M. Henry, and B. Murr. *Science* **125**, 350 (1957).
21. R. Paschke, R. Chang, and D. Yaung, *Science* **125**, 881 (1957).
22. C. Palm and M. Calvin, *J. Am. Chem. Soc.* **84**, 2115 (1962).
23. S. Fox, J. Johnson, and A. Vegotsky. *Science* **124**, 923 (1956).
24. J. Oro, A. Kimbell, R. Fritz, and F. Master. *Arch. Biochem. Biophys.* **85**, 115 (1959); J. Oro and S. Kamat. *Nature* **190**, 442 (1961).
25. C. Lowe, M. Ress, and R. Markham. *Nature* **199**, 219 (1963).

26. K. Harada and S. W. Fox. *Nature* **201**, 335 (1964); *In* "The Origin of Prebiological Systems" (S. W. Fox, ed.). Academic Press, New York, 1965.
27. W. Groth. *Angew. Chem.* **69**, 681 (1957); W. W. Groth and H. Weyssenhoff, *Ann. Physik* [7] **4**, 69 (1959).
28. A. Terenin. *In* "Origin of Life on Earth" (A. I. Oparin, ed.). Pergamon Press, Oxford, 1959.
29. N. Dodonova and A. Sidorova. *Biofizika* **6**, 149 (1961).
30. T. Pavlovskaya, A. Pasynskii, and A. Grebsinkova. *Dokl. Akad. Nauk SSSR* **135**, 743 (1960).
31. A. Deschreider. *Nature* **182**, 528 (1958); R. Cultrera and G. Ferrarri. *Agrochimica* **5**, 108 (1961).
32. G. Grieppe, and M. Galotti, *Gazz. Chim. Ital.* **59**, 507 (1929); H. Dodonova and A. Sidorova. *Biofizika* **7**, 31 (1962).
33. E. E. Ellenbogen. *Abstr. 134th Meeting Am. Chem. Soc., Chicago,* 1958 p. 47; O. Petri, K. Bakhadur, and N. Potan. *Biokhimiya* **27**, 708 (1962); K. Otozai, S. Kume, S. Nagai, T. Jamamoto, and S. Fukushima. *Bull. Chem. Soc. Japan* **27**, 476 (1954).
34. N. Kuhn and E. Braun. *Naturwissenschaffen* **17**, 227 (1929).
35. A. Byk, *Z. Physik. Chem.* **49**, 61 (1904); *Naturwissenschaffen* **13**, 17 (1925).
36. J. Bernal. "The Physical Basis of Life." Routledge, London, 1951; *Usp. Khim.* **19**, 401 (1950).
37. A. Terent'ev and E. Klabunovskii. *In* "Origin of Life on Earth" (A. I. Oparin, ed.). Pergamon Press, Oxford, 1959.
38. G. Wald. *Ann. N.Y. Acad. Sci.* **69**, 352 (1957); G. Eiring, L. Dzhons, and Dzh. Spaiks. *In* "Gorizonty biokhimii" ("Horizons of Biochemistry"), p. 174. Izd. "Mir," 1964.
39. A. Pasynskii. *In* "Origin of Life on Earth" (A. I. Oparin, ed.). Pergamon Press, Oxford, 1959.
40. J. Oro. *Biochem. Biophys. Res. Commun.* **2**, 407 (1960); J. Oro and A. Kimball. *Arch. Biochem. Biophys.* **94**, 217 (1961); **96**, 293 (1962).
41. J. Oro. *Nature* **191**, 1193 (1961).
42. J. Oro. *In* "The Origin of Prebiological Systems" (S. W. Fox, ed.). Academic Press, New York, 1965.
43. S. Fox and K. Harada. *Science* **133**, 1923 (1961).
44. J. Oro. *In* "Problemy evolyutsion. i tekhnich. biokhimii" ("Problems of Evolutionary and Technical Biochemistry"), p. 63. Izd. "Nauka," 1964.
45. E. Mariani and G. Torraca, *Intern. Sugar J.* **55**, 309 (1953).
46. J. Oro. *Federation Proc.* **25**, 80 (1962).
47. M. Calvin. "Chemical Evolution," London Lectures, 1961; C. Palm and M. Calvin. *Univ. Calif. Radiation Lab. Rept.* (1961).
48. C. Palm and M. Calvin. Submitted to *J. Am. Chem. Soc.* (1961).
49. C. Ponnamperuma, R. Lemmon, R. Mariner, and M. Calvin, *Proc. Natl. Acad. Sci. U.S.* **49**, 737 (1963).
50. B. Pullman and A. Pullman. *Nature* **196**, 1137 (1962). *In* "Molecular Orbitals in Chemistry, Physics, and Biology" (P.O. Löwden and B. Pullman, eds.), p. 547. Academic Press, New York, 1964.

51. C. Ponnaperuma. **NASA** (1964).
52. S. Akabori. *Kagaku Tokyo* **25**, 54 (1955).
53. S. Akabori. *In* "Origin of Life on Earth" (A.I. Oparin, ed.). Pergamon Press, Oxford, 1959.
54. A. Wilson. *Nature* **188**, 1007 (1960).
55. S. Fox, K. Harada, and A. Vegotsky. *Experientia* **15**, 81 (1959).
56. I. Kulaev. *Rept. 1st All-Union Biochem. Congr., Leningrad, 1964* Abstracts, No. 1, p. 12. Izd. Akad. Nauk SSSR, 1964.
57. I. Kulaev and V. Vagabov. *Biokhimiya* (1965).
58. M. Szymona and O. Szymona. *Bull. Acad. Polon. Sci., Ser. Sci. Biol.* **9**, 371 (1961).
59. A. Belozerskii. *In* "Origin of Life on Earth" (A. I. Oparin, ed.). Pergamon Press, Oxford, 1959.
60. G. Schramm, H. Grotsch, and W. Pollman. *Angew. Chem.* **74**, 53 (1962).
61. G. Schramm. *In* "The Origins of Prebiological Systems" (S. W. Fox, ed.), Academic Press, New York, 1965.
62. A. Schwartz, E. Bradley, and S. W. Fox. *In* "The Origins of Prebiological Systems" (S. W. Fox, ed.). Academic Press, New York, 1965.
63. C. Ponnamperuma, C. Sagan, and R. Mariner. *Nature* **199**, 222 (1963).
64. C. Ponnamperuma. *In* "The Origins of Prebiological Systems" (S. W. Fox, ed.). Academic Press, New York, 1965.
65. "Advances in Organic Geochemistry" (E. Ingerson, ed.), Monograph No. 15, Earth Sci. Ser. Pergamon Press, Oxford, 1964.
66. E. Krinov. "Osnovy meteoritiki" ("Foundations of Meteoritics"). Gosteoretizdat, 1955.
67. G. Berzelius. *Pogg. Ann.* **33**, 115 and 143, (1834).
68. F. Wöhler, *Sitzber. W. Akad.* **41**, (1859).
69. E. Cohen. "Meteoritenkunde." H. J. Stuttgart, 1894.
70. G. Mueller. *Geochim. Cosmochim. Acta* **4**, No. 1/2 (1953).
71. A. Wilson, *Nature* **196**, 11 (1962).
71a. A. Gurstein. *Priroda* **6**, 3 (1967).
72. K. Sztrokay, V. Tolmay, and M. Foldvari. *Vogl. Acte Geol.* **7**, 1 (1961).
73. G. Vdovykin. *Geiokhimya* No. 2 (1962).
74. B. Nagi, W. Meinschein, and D. Hennessy. *Ann. N.Y. Acad. Sci.* **93**, No. 2 (1961); *Nature* **189**, 967 (1961).
75. C. Sagan, *Natl. Acad. Sci. – Natl. Res. Council, Publ.* **757**, (1961).
76. J. Yames. *Sci. Am.* July (1963).
77. G. Fesenkov. *In* "Problemy evolyutsion. i tekhnich. biokhimii" ("Problems of Evolutionary and Technical Biochemistry"), p. 5. 1964.
78. D. Martynov. *In* "Vozniknovenie zhiznii vo Vselennoi" ("Origin of Life in the Universe"). p. 56. Izd. Akad. Nauk SSSR, 1963.
79. W. Sinton. *Science* **130**, 1234 (1959).
80. N. Colthup. *Science* **134**, 529 (1961).
81. D. Rea. *Nature* **200**, 114 (1963); D. Rea, T. Belsky, and M. Calvin, *Science* **141**, 923 (1963).
82. R. Young, C. Ponnamperuma, and B. McCaw. *Repts. COSPAR Conf. Florence,*

1964 Abstracts.
83. D. Hull. *Nature* **186**, 692 (1960).
84. H. Urey. "The Planets, Their Origin and Development." Yale Univ. Press, New Haven, Connecticut, 1952.
85. C. Sagan. *Radiation Res.* **15**, 174 (1961).
86. A. Szutka, J. Hazel, and W. McNabb. *Radiation Res.* **5**, 597 (1959); A. Szutka. *ibid.* **19**, 183 (1963); *In* "The Origins of Prebiological Systems" (S. W. Fox, ed.). Academic Press, New York, 1965.
87. A. Krasnovskii and A. Umrikhina. *Dokl. Akad. Nauk SSSR* **155**, 691 (1964).
88. M. Dolin. *Bacteria* **2**, 425 (1961); A. Oparin, E. Kharat'yan, and N. Gel'man. *Dokl. Akad. Nauk SSSR* **157**, 207 (1964).
89. C. Hinshelwood. "The Chemical Kinetics of the Bacterial Cell." Oxford Univ. Press (Clarendon), London and New York, 1947; *J. Chem. Soc.* p. 745 (1952).
90. M. Dixon and E. C. Webb. "Enzymes." Academic Press, New York, 1964.
91. Simplicius on Aristotle's De Caelo, 586, 29. H. Diels. "Die Fragmente der Vorsokratiker," 2nd ed., Vol. 1, p. 190 (Empedocles, 1357). Berlin, 1906.
92. J. Haldane. *In* "The Origins of Prebiological Systems" (S. W. Fox, ed.). Academic Press, New York, 1965.
93. H. Pattee. *Biophys. J.* **1**, 683 (1961); B. Vanyushin. *Priroda* April, p. 29 (1963).
94. A. Oparin and K. Serebrovskaya. *Dokl. Akad. Nauk SSSR* **148**, 943 (1963).

CHAPTER 4

ORIGIN OF PREBIOLOGICAL SYSTEMS

It is characteristic of life that it is not simply dispersed in space, but is set off from the outer world into individual, highly complicated systems — organisms. They could have been formed only on the basis of a length evolution, through the gradual perfection of some far simpler original systems isolated from the general homogeneous solution of organic compounds (1).

Initially, such systems could have been simply isolated portions of the "primitive soup" separated from the solution as a whole by some sort of boundary, but able to interact with the environment.

As an example of such a system, readily formed even in present-day nature, can be cited the "bubbles" of Goldacre (2) (Fig. 10). He recently described the formation on the surface of natural bodies of water of small, closed bubbles, contained within a protein-lipoid sheath. They arise simply from the effect of the wind on the protein-lipoid film which lies on the water surface. At present, this film is formed from the decomposition products of dead organisms, but Wilson (3) showed experimentally that it could also have been formed on the primeval ocean from abiogenic carbon compounds.

As stated in the last chapter, Fox and his co-workers (4), simulating conditions in a volcanic eruption, heated a mixture of 18 amino acids (using a surplus of the dicarboxylic acids) to 170°C in a block of lava. Proteinoid polymers were thereby formed, which in aqueous solutions formed multimolecular microspheres $2\text{-}7\mu$ in diameter, in their external appearance resembling the structural elements which have been found in meteorites (Fig. 11). The synthetic proteinoid microspheres are able to keep their shape for a long time and can withstand centrifugation at 3000 rpm. They are stable to operations involved in the preparation of ultrathin sections, as first demonstrated by Young and Munoz (5); they are capable of taking a gram stain, and exhibit several other qualities

101

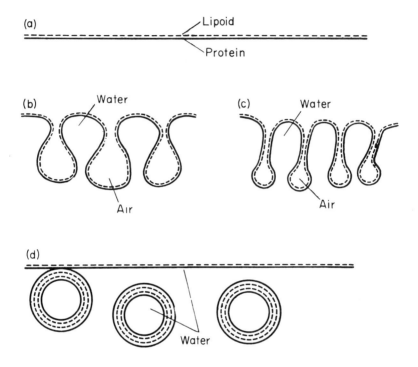

FIG. 10. Formation of small bubbles (After R. Goldacre). a–d, Successive stages.

resembling those of bacterial cells (6). The electron microphotographs which were shown at the conference at Wakulla Springs in 1963 by Fox and by Young (7) in fact are strikingly similar to corresponding photographs of bacterial sections. A granular structure is visible in the microspheres, and even a double membrane, blanketing these artificial formations at the surface like the protein-lipoid membrane of living cells (Fig. 12).

All this demonstrates a remarkable capacity for extensive self-organization of macromolecular complexes formed under near-primitive conditions, and must be taken into account in a study of the formation of protoplasmic structures of primeval organisms.

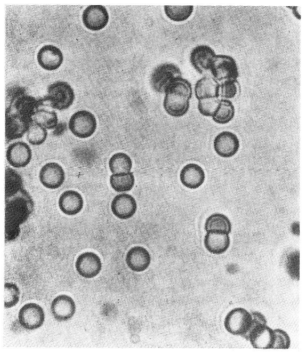

FIG. 11. The simple microspheres of S. Fox under an optical microscope (diameter 7μ).

However, we must always think of any system as organized not only in space but also in time, since these two aspects of organization are inseparable. On one hand, any system has certain dimensions and a structure conforming to the interarrangement of its parts, and on the other hand, it has a definite combination and sequence of the processes occurring in it. For protoplasm, the latter form of organization, continuously performing metabolic processes, is especially significant, since even the protoplasmic structure itself is not static but dynamic in character. Its stability is determined not by its immutability, but on the contrary by constantly recurring protoplasmic biochemical processes, regulated by a combination of synthesis and decomposition (8).

These comparatively simple yet abiogenic formations which once were made up of organic material in the waters of the original ocean

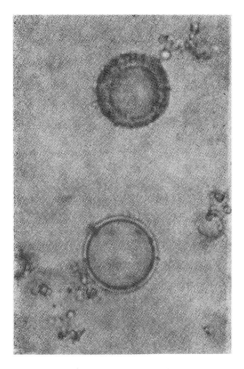

FIG. 12. The same microspheres as in Fig. 11 under the electron microscope. A double membrane is visible.

and which were initial systems on the pathway to the origin of life, must have evolved both with respect to spatial organization in the direction of increasing complexity and the perfection of structure, and also with respect to the organization of a definite sequence of processes at the time when the system was being created.

The establishment of these evolutionary pathways is of the utmost importance to investigators working on the problem of the origin of life; the evolution in which a definite order of metabolic processes has arisen from the chaos of disordered and tangled reactions of the "primitive soup." The solution of this exceptionally complex task must be approached from two sides. On one side, an extensive comparative biochemical investigation of modern organisms is needed, enabling us to develop those reactions and combinations of reactions which are the groundwork of metabolic processes in absolutely all living things, and which thus can be regarded as the oldest original metabolic systems arising at a time when the tree of life was still not divided into its separate branches.

The other method of investigation consists in simulating experimentally in some measure those phenomena which occurred in the "primitive soup." The goal of these experiments consists in demonstrating the possibility (or even the necessity) of the emergence in these conditions of some sort of individual systems which in their evolutionary development gradually organized the combinations of reactions occurring in the system, bringing them successively closer to an indigenous, even though most primitive, metabolic process.

Fox's microspheres, since they are obtained thermally, do not present very promising results from this point of view. Their structure is static. This is very advantageous for electron microscopic investigations, but creates many difficulties when it comes to converting them into dynamic systems which could be used for modeling the evolution of metabolism.

A characteristic feature of biological polymers — proteins, nucleic acids, carbohydrates, and lipoid polymers — is their clearly expressed ability to form complexes with other high molecular weight organic substances. The formations emerging from these processes often have physical and chemical properties which differ significantly from the properties of the individual substances composing them.

Bungenberg de Yong (9), observed that dilute solutions of gelatin and gum arabic on mixing became turbid at certain acidities and temperatures quite normal in nature. This was due to the fact that the molecules of gelatin and gum uniformly distributed throughout the whole volume of the solution began to unite into integral molecular "clusters" or "mounds." When the dimensions of these clusters reached a certain quite considerable size, they separated from the solution in the form of drops visible under the ordinary optical microscope; the drops were floating freely in the water surrounding them, which now was almost entirely deficient in these polymers (Fig. 13).

Bungenberg de Yong (10) called the formations which he obtained coacervate drops. He showed (11) that the process of coacervation is one of the most powerful methods of concentrating highly polymerized substances from very dilute solutions. Thus, for example, coacervate

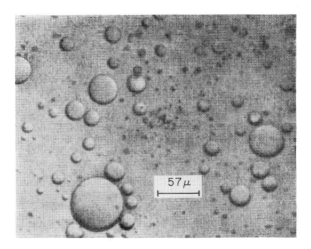

FIG. 13. Coacervate droplets of gelatin and gum arabic.

drops have been isolated from a solution containing gelatin at a dilution of 0.001%. The polymer concentration in the drops themselves reached several times 10%. Although the drops are liquid in consistency and in several cases are hydrophilic, they are sharply segregated from the aqueous solution surrounding them (the so-called equilibrium liquid) by a boundary surface, although they can react chemically with the external medium. Both Bungenberg himself (12) and subsequent authors (13) obtained coacervate drops not only from gelatin and gum arabic but also from many other organic polymers. Specifically, in our laboratory at the Bakh Institute of Biochemistry, Serebrovskaya (14) obtained 17 forms of coacervates of varying composition, in drops which combined various proteins (among them albumin and histone), nucleic acids, polyglucosides, lipoids, chlorophyll, etc. (Fig. 14).

In all these cases, substances were used which had a biogenic origin and were isolated from living organisms. This of course was because it was simplest to obtain the needed polymers in the laboratory in this way. However, for this reason, the opinion was even expressed in the literature that only substances of a biological origin with their ordered secondary and tertiary structure can form coacervate drops (15). We were able to refute this opinion experimentally and showed that the degree of polymerization of the substances combining in drops is the decisive element in coacervation, and not the ordered arrangement of

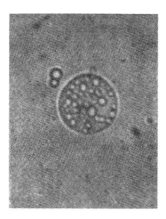

FIG. 14. Nucleoprotein coacervate, RNA + histone, pH 9 (X 800).

the monomers in their polymeric chain (16). For this purpose we synthesized polyadenine *in vitro* from adenosine diphosphate in aqueous medium (Fig. 15) in the presence of, or in parallel with, the formation of another polymer (polylysine, for example). In this process, as soon as a certain degree of polymerization of the synthesized substance is attained, coacervate drops containing these polymers separate. Consequently, coacervate drops are formed not only by mixing solutions of natural proteins, nucleic acids, and other biogenic compounds, but also by solutions of synthetic polymers with a monotonic chain structure or polymers lacking a complex ordered secondary structure. A similar

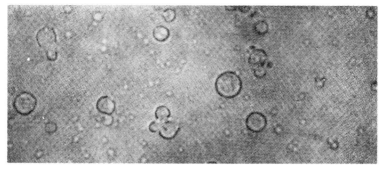

FIG. 15. Coacervate droplets, formed during the synthesis of polyadenine in the presence of histone.

TABLE 5
Content of Dry Material in Coacervate Drops

No. expt. *	Diameter (10^{-4} cm)	Volume (10^{-12} cm^3)	Weight (10^{-12} gm)	Concentration (C) in %	(C) Drops (C) Solution
		Protein−carbohydrate			
		⟨ Serum albumin−gum arabic (pH 4.40) ⟩			
1	3.08	15.3	5.2	34	50
2	3.30	18.8	6.0	32	47
3	6.16	122	12.9	21	31
		Histone−gum arabic (pH 5.5−6)			
4	3.30	18.8	9.7	52	94
5	6.16	122.1	53	43	79
6	8.36	305.3	102.3	38	61
		Salmon protamine−gum arabic (pH 6.0)			
7	4.40	44.6	13.8	31	39
8	4.62	51.5	15.5	30	37
9	8.69	342	82	24	30
		Sturgeon protamine−gum arabic (pH 5.8−6.0)			
10	3.52	22.8	11.3	50	33
11	6.27	127.9	44.1	34	23
12	7.48	218.6	63.4	29	19
		Clupein−gum arabic (pH 6.0−6.5)			
13	1.98	4.1	1.5	36	37
14	3.19	16.8	5.4	32	33
15	6.27	128.2	35.9	28	29
		Protein−protein			
		Histone−serum albumin (pH 5.5−6.0)			
16	2.20	5.6	2.4	44	44
17	3.96	32.4	11.3	35	35
18	5.50	86.9	15.6	18	18
		Histone−gelatin (pH 6.2−7.3)			
19	4.18	38.2	13.5	35	60
20	5.50	86.9	25.7	30	50
21	11.66	762.2	107.2	15	25
		Clupein−gelatin (pH 8.6−8.8)			
22	3.30	18.8	2.6	14	13
23	5.61	91.8	10.1	11	10
24	7.92	259.6	20.8	8	7
		Protein−nucleic acid			
		Clupein−RNA (pH 8.0−8.6)			
25	1.98	4.1	2.4	59	348
26	4.84	59.3	26.0	44	258
27	7.15	191.3	58.4	31	179

TABLE 5 *(Continued)*
Content of Dry Material in Coacervate Drops

No. expt.	Diameter (10^{-4}cm)	Volume (10^{-12}cm^3)	Weight (10^{-12}gm)	Concentration (C) in %	(C) Drops (C) Solution
		Clupein—DNA (pH 7.6—8.2)			
28	1.76	2.8	2.3	56	431
29	4.04	35.3	9.1	26	185
30	9.24	412.9	56.2	14	85
		Histone—RNA (pH 7.0—7.6)			
31	1.76	2.8	1.5	58	116
32	4.18	38.3	5.6	15	29
33	15.84	2 080.1	100	4	8
		Histone—DNA (pH 6.8—7.6)			
34	2.2	5.6	2.3	42	84
35	4.4	44.6	11.4	25	51
36	16.72	2 446.5	169	7	14
		Protein—carbohydrate—nucleic acid			
		Gelatin—gum arabic—DNA (pH 3.8—4)			
37	2.32	6.5	2.4	37	55
38	5.35	79.4	19.8	25	37
39	6.43	137.90	31.6	23	34
		Gelatin—gum arabic—DNA (pH 3.8—4)			
40	34.5	21 400	4 800	22	33
41	41.4	37 100	5 800	16	23
42	62.1	125 100	12 800	10	15
		Gelatin—gum arabic—RNA (pH 3.8—4)			
43	34.5	21 400	42 800	20	30
44	53.8	81 300	11 200	14	21
45	179.4	3 016 800	235 010	8	11
		Protein—carbohydrate RNA			
46	29.9	14 000	1 200	8	12
47	85.1	322 000	19 900	6	9
48	163.3	2 275 300	23 200	1	1
		Protein—lipid			
		Gelatin—oleate (pH 8.4—8.6)			
49	7.7	238.9	55	28	3.6
50	17.78	2 941.8	505.4	17	2.7
51	24.36	7 565.9	1 190.2	16	2.4
		Multi—component coacervate			
		Phosphorylase—histone—starch—gum arabic etc. (pH 6.0—6.2)			
52	2.42	7.4	3.7	50	75
53	4.18	38.2	14.2	37	55
54	5.50	86.9	24.1	28	41

separation of coacervate drops must have taken place in the "primitive soup" where nonspecific polymerization of a variety of organic compounds occurred, amino acids and mononucleotides in particular. In addition to this, coacervate drops are the sort of individual, multimolecular systems, which can easily shift from a stable to a dynamic state through the constant interaction of these formations with their environment in the manner of open systems. Thus coacervate drops are very convenient models for the reproduction in the laboratory of those tentative pathways which both the structures and the metabolism of the original systems followed toward the origin of living systems (17).

Coacervate drops can be formed from two or more components. The size of the drops and their concentration of polymers depends first of all on the nature of the substances composing them. But even drops of the same composition can vary greatly in their volume and weight. We present a table (Table 5, p. 108) derived from the work of Evreinova (18) which gives the volume, weight of the drop, polymer concentration in individual drops (C), as well as the ratio of this concentration to the concentration of the same substances in the surrounding solution.

It is evident from this table that when coacervation occurs the polymers are concentrated in the drops by many orders of ten and even sometimes by hundreds of times over the concentration in the solution. Drops of the same composition can differ greatly in the degree to which the polymers are concentrated in them, but the less the volume of a given individual drop, the higher the total concentration of material in it, and the less the hydration in the drop.

Using interference microscopy, Evreinova demonstrated a considerable variation in the concentration of material not only in the individual drops of the same composition but even within the limits of the same drop (19) (Fig. 16). Vacuoles can occur in which the polymer content is sharply decreased. In contrast, in other cases some structure formation in separate parts of the drop can be observed. In the presence of such structures some coacervate formations occur which are not always spherical.

Under an ordinary optical microscope it is quite difficult to find any structure in the coacervate drops, but it is apparent to a certain degree in photographs of the drops (without preliminary drying) under the electron microscope provided with the gas microchamber system of Stoyanova (20) (Fig. 17).

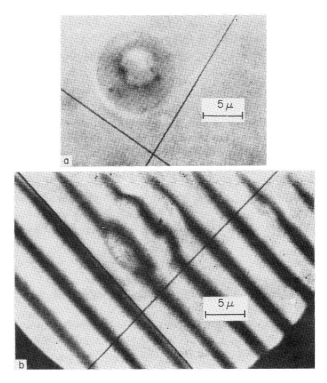

FIG. 16. a. Coacervate droplet with a vacuole. b. The same droplet under the interference microscope.

The non-uniform distribution of material in the drops is particularly evident when one of the components is a nucleic acid and the other a histone. The nucleic acid concentration in individual structural formations of coacervate drops was measured by Evreinova (21) by means of an MUF-4 ultraviolet microscope (Fig. 18). She showed by this method that the concentration can vary within wide limits, revealing the unique characteristics of individual drops even when they come from the same solution. According to the data of Bungenberg de Yong (22) and several subsequent authors (23), films can form on the surface of coacervate drops in the presence of lipoids. But even in the absence of such films, there always exists a sharp boundary between the drops and the surrounding medium. Nevertheless, the drops are not isolated from this

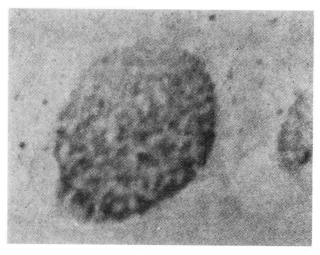

FIG.17. Coacervate droplet of serum albumin + gum arabic + RNA + RNase, (× 1200). Gas chamber of the electron microscope.

medium, but are capable of interacting with it; this is very important for their future evolution.

First of all, this interaction is expressed in the ability of the drop to absorb material selectively from the external medium (equilibrium liquid). It is very easy to demonstrate this ability by adding various dyes to the equilibrium liquid. In this way, we can observe directly under the microscope how the dye gradually concentrates in the drops and the surrounding solution correspondingly becomes lighter in color. The measurements of Evreinova have shown that such dyes as neutral red, methylene blue, etc. are concentrated in the drops (consisting, for example, of gelatin and gum arabic) by an amount many factors of ten more than their initial concentration in the starting solution. Evreinova (24) and co-workers also demonstrated the selective concentration in coacervate drops of amino acids; however, while the tyrosine concentration, for example, exceeded the concentration in the surrounding solution by 100 times, the concentration of others (tryptophan, for example) was only doubled. In contrast to this, sugars and mononucleotides were equally distributed between the drops and the surrounding solution, so that equal volumes of drops and solution contained an equal

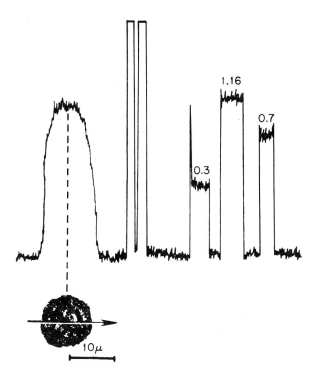

FIG. 18. Quantitative determination of the nucleic acids in a coacervate droplet (clupein + DNA).

amount of these substances. Of course, the ability to concentrate some compounds and the magnitude of this concentration is very greatly altered by changes in composition and structure of the drops.

However, the ability to concentrate material from the surrounding solution still does not make the coacervate drop a model of a living system. Equilibrium between the drop and the surrounding solution is established very quickly; it reaches a thermodynamically stable state and becomes a static system. In contrast to this, any cell or multicellular organism during the course of its existence is in a state of continuous interaction with the surrounding environment. The living system exists only as long as it is a continuous cycle and the numerous biochemical processes involved with its metabolism proceed at a rapid rate. As soon as these processes are broken off, or basically altered, the protoplasmic system itself is destroyed. Its prolonged existence and the constancy of its form are associated not with unchangeability, not with rest, but with constancy of motion. Thus protoplasm is not a static but a stationary system (25).

The great ancient Greek dialectician Heraclitus (26) observed this characteristic feature of living beings; he taught that as matter flows through our bodies, as in a brook, it is being renewed just as is water in the stream. It is the analogy of this stream, or simply the current of water flowing from a water tap which enables us to understand in their simplest form the most important organizational aspects of irreversible or open systems, specifically those which exist in living protoplasm. If the tap is slightly open and the pressure in the water line remains constant, the stream of flowing water maintains its external form almost unchanged, its "frozen" form as it were. But we know that this form is only the visible representation of the continuous flow of water particles which constantly enter and leave it at a definite rate. The significance of this system consists in the fact that ever-fresh particles of water are uniformly rushing through the stream all the time. If we stop this process, the stream itself as a definite system disappears.

Analagous to this, and at the basis of the organization of protoplasm, is that stationary state in which the living organism is constantly exchanging matter and energy with the medium surrounding it; in this state a series of irreversible coupled processes are occurring all the time at a definite constant rate, as a result of which the matter entering from the external medium undergoes several transformations in the organism, and the decomposition products are again ejected into the external medium.

The use of labeled atoms in biochemical and physiological investigations (27) has definitely proved that all the matter in a living body, its proteins, nucleic acids, lipoids, etc. is completely renewed within a very short interval of time, that the material substrate of life is in a constant process of exchange with the surrounding medium, constantly being decomposed and again synthesized from the matter of the external world. Thus, the principle of the harmony of organism and environment has been completely confirmed; the principle that the living being must never be regarded as isolated from the environment, as outside of this harmony.

Prigogine (28) in his interesting book "Introduction to the Thermo-dynamics of Irreversible Processes" divides all finite systems into the following basic groups: open, closed, and isolated. To the first group belong systems in which matter and energy are constantly being ex-changed with the environment. In closed systems this exchange is limit-ed to energy alone, and matter is not exchanged. To the third group belong systems completely isolated from their environment, in which neither energy nor matter are exchanged. The latter two groups of systems can be combined under the general heading of closed systems, as distinguished from open systems, to which living organisms belong.

In closed systems, only matter included in the system can participate in a chemical reaction. The constancy of properties of the system with time is characterized by an equilibrium state, in which the rate of a reaction going in one direction is equal to the rate of the reverse reaction. The thermodynamic criteria of this equilibrium are the minimal significance of the free energy of the system and the maximum signifi-cance of its entropy (in other words, its transition to a more probable state from all possible states). The processes spontaneously occurring in a closed system cannot convert it into a less probable state, that is, they can only keep the entropy of the system constant, or increase it, depending on whether we are dealing with a reversible or irreversible process. When the entropy of the system increases, equilibrium is not maintained and on the contrary, when equilibrium is approached, the rate of increase of entropy is equal to zero. In contrast to this, in the open system, matter from the external medium is continuously entering into a system segregated from the medium and chemical compounds arising from reactions in the system are being removed. Thus the constancy of properties of such open systems with time is characterized not by thermodynamic equilibrium (as was observed in closed systems) but by the onset of a stationary state during which a constant rate of one-way chemical changes and diffusion of the substances into the system is maintained.

Thermodynamic equilibrium and the stationary state resemble each other in the sense that in these cases the systems keep their properties constant with time, but they differ radically in that no change in free energy generally occurs in equilibrium ($dF = O$), while in the stationary state it is changing continuously but at a constant rate ($dF = \text{const}$).

We can give the following elementary example as an illustration. A simple bucket of water can represent the closed static system which keeps its water level constant because no action is occurring here. On the other hand, a tank into which water flows in continuously through a pipe at one end and out the other, is a stationary open system. The water level in the tank can remain unchanged, but only if there is a definite constant relationship between the rates of inflow and outflow. By changing these relationships we can create any other level, which then on the same basis will keep its stationary state.

In the very simple example given, we have chosen a system in which no chemical reactions occur.

However, from the point of view of understanding vital processes, we are much more interested in the chemically open system. Matter is continuously flowing in from the external medium and being kept from contact with the medium in some sort of system; this matter, however, undergoes chemical changes, and the reaction products thereby produced are returned to the external medium. Thus the stability of this system with time is characterized by a stationary state in which not only the rates of inflow and outflow of matter into the system are balanced, but also the rates of the chemical changes occurring in the system.

Since there is a constant rate of diffusion and reaction, the system reaches a stationary state, the components of the system reaching a certain level. With any change in these parameters the "equilibrium" of the system is disturbed, but then a new stationary state is established; there is no limit to the number of possible states. If, therefore, a catalyst which speeds up a certain reaction is introduced into the system, the levels of the components can be changed; this action is forbidden in a closed system in which a catalyst can change only the rate at which equilibrium is attained, but not its state (29).

In the living cell we are dealing with an incomparably more complex system than the extremely simple chemical diagram which we have given. Here, in the first place, it is not a single individual chemical reaction that has significance, but a whole chain of chemical reactions, all closely interconnected. This chain can be linear and unbranched, but it can also be branched or even closed into a cycle. The chemical processes in these cycles recur in a certain pattern. However, in individual links of the chain, irreversible branching reactions always occur, causing biological metabolism as a whole always to flow in a single direction.

In protoplasm many chains and cycles of reactions are united in a single, very highly branched metabolic network constructed in a regular pattern which Hinshelwood (30) compares with a well-developed railroad network, on which many trains move simultaneously at different speeds.

A simple solution or an homogeneous mixture of organic compounds, from this point of view, represents a very broad and completely extravagant range of chemical possibilities. In this area we can move in any direction with equally great hindrance and thus also with an equally small rate. In contrast to this, in protoplasm a definite route for biochemical processes has been laid down, a whole network of "efficiently built lines," along which chemical reactions and the energy conversions connected with it proceed uninterruptedly with colossal speed and according to a strictly observed schedule, while at the basis of this whole organization of protoplasm with time is the relationship of the rates of interconnected metabolic reactions.

In order for the coacervate drop to approach living things in its organization, in order for it to serve to a certain degree, even though at the least complex level, as a model of these living things, it must be converted from the static state to the stationary, its interaction with the external medium must "flow" like a stream. And for this it is necessary not only that the matter in solution surrounding the drop by selectively absorbed by it, but also that it undergo some chemical change in the drop.

The relationships resulting from this can be shown in the following diagram:

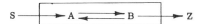

where the rectangle represents the system (for example the coacervate drop), S and Z the external medium, A the matter entering the drop, B the reaction product capable of diffusing into the external medium.

If the reaction $A \rightarrow B$ goes more rapidly than the reverse reaction, if it proceeds in the system at a greater rate than in the surrounding medium, the concentration of A in the system will fall and this will destroy the equilibrium between it and the medium, making it possible for increasing quantities of A to enter the system. In contrast to this, the concentration of B in the reaction process increases, and this causes the passage of B into the external medium. As a result of this a constant and unidirectional flow of matter through the system can occur, dF thereby remains constant, being replaced at the expense of energy flowing with A from the external medium and being released in the system in the course of the reaction $A \rightarrow B$ (of course, only if this reaction is exergonic).

The most effective way of increasing the rate of such a reaction in the system is by the introduction of a corresponding catalyst, which can either be an organic compound or an inorganic salt or a combination of both. It is not necessary at all that this catalyst be initially formed in the system itself, it can partially or completely enter the system from the external medium, but it must be accumulated here like the adsorbed dye was accumulated in the coacervate drop.

A somewhat more complex scheme is obtained when there are not one but two interconnected reactions:

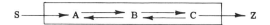

In this case, depending on the relationship between the rate of reactions $A \rightarrow B$ and $B \rightarrow C$, B can accumulate in the system or rapidly disappear from it; if B is a polymer composing part of the system, it can increase or decrease in its volume and weight.

In order to reproduce these phenomena in model experiments we (31) used coacervate drops as a system interacting with the external medium. In the majority of cases solutions of natural polymers (polysaccharides, proteins, nucleic acids, etc.) isolated from various organisms are mixed to form coacervate drops. Of course, we allowed for the fact that since these substances have a rigidly determined intramolecular structure, they could not have been in the "primitive soup" of the Earth's atmosphere; however it was easier with this material to establish those phenomena which are also characteristic of the nonstructured polymer (which was later confirmed by special experiments).

Similarly, we used an enzyme preparation as a catalyst and not simpler (but less effective) accelerators, since it permitted us to run our experiments rapidly within a period suited to laboratory conditions (32). Later we propose to substitute for the enzymes a less complete organic or inorganic catalyst. But at the present beginning stage of the investigation, the use of natural polymers and enzymes gave an invaluable advantage.

If a preparation of potato phosphorylase (glucosyl transferase) is added to a coacervate formed by mixing solutions of gum arabic and histone at pH 6–6.2, the enzyme will be almost entirely concentrated in the drops. When glucose 1-phosphate is dissolved in the equilibrium liquid, starch, which can easily be detected by the iodine test, begins to accumulate in the drops. The increase in starch in a single coacervate drop can be measured directly by interference microscopy. As soon as 30 minutes later its weight increases by 50% and its volume more than 1 ½ times (33).

Since glucose 1-phosphate is not significantly adsorbed by the coacervate drops mentioned above, its concentration in drops and in the surrounding solution always remains approximately the same. Thus, in order for such a large amount of starch to be formed in the drop, more and more glucose 1-phosphate must enter the drop from the surrounding medium as it is used up in the synthesis of the polymer. When β-amylase is included in the drop in addition to the phosphorylase, the starch formed in the drop breaks down into maltose which passes into the external medium where it can be tested for and measured.

Thus the whole process of synthesis and decomposition which we have described can be illustrated in the following diagram of the flow through a coacervate drop (represented here by a rectangle).

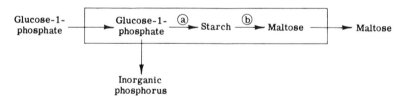

Depending on the ratio between the rates of reactions a and b, the amount of polymer (starch) formed in the drop from the material in the external medium can increase or decrease, the drop can grow or be broken up. To a certain degree this system can serve as a very simple model of the flow of material through a living cell.

In a similar manner we have produced both enzymic decomposition (34) and enzymic synthesis of polynucleotides (35) which were one of the components of our coacervate drops. The decomposition was accomplished by means of ribonuclease incorporated in a drop consisting of RNA, serum albumin, and gum arabic. As a result, the RNA was decomposed within the drop and its products (mononucleotides) were released into the external medium. This could be established by chem-

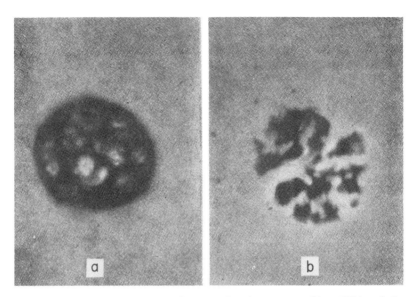

FIG. 19. Coacervate droplet of serum albumin + gum arabic + RNA and the enzyme ribonuclease under the electron microscope. a. Before the beginning of breakdown. b. After a 15-minute breakdown period.

ical analysis, as well as seen directly in photographs obtained in the gas chamber of the electron microscope in the beginning and at the end of the enzymic action (Fig. 19).

The reverse process, enzymic synthesis of polynucleotide in coacervate drops from mononucleotides soluble in the external medium, we accomplished by means of a bacterial polynucleophosphorylase, incorporated in a drop of histone and RNA. Adenosine diphosphate (ADP) served as a substrate, and polyadenine accumulated in the drops. Even in the absence of RNA this substance forms coacervates with histone, and in the drops thus formed the polymer synthesis goes incomparably faster and more intensively than in a homogeneous solution of substrate and enzyme (36).

Thus we now have dynamically stable coacervate drops, interacting with the external medium in the manner of an open system; drops which are capable not only of remaining in a stable state, but also of increasing their volume — growing by an accelerated polymerization of material from the external medium. Such drops can serve as models of those complex multimolecular systems which must have been isolated from the aqueous solution of the "primitive soup," and which were formed from the polypeptides, polynucleotides, and other polymers present there, without, however, any definite sequence of monomeric residues in their chains.

It is true that the growth of such models can take place only when energy-rich phosphorus compounds are present in the external medium. But as the experiments of Ponnamperuma *et al.* have indicated, such compounds must have been synthesized abiogenically in the "primitive soup," particularly through the mediation of energy from short-wave ultraviolet light. Consequently their presence in the pre-actualistic reducing hydrosphere of the Earth is by no means excluded, but it is unlikely that the amount produced would be sufficient to sustain the steady existence of coacervate drops for any length of time. After some beginning period, only those drops could be longer maintained as stationary systems which possessed the ability to form high energy compounds by processes leading to the release of free energy.

In the reducing conditions of the "primitive soup," the most likely of these processes would be oxidation-reduction reactions involving the transfer of hydrogen or electrons.

Apparently, it is precisely reactions of this type which were the early links in the nascent metabolic process. This is indicated by their astonishing universality, their obligatory participation in all living processes known to us, in absolutely all living things. No wonder that Kluyver (37) on the basis of very extensive comparative biochemical data rightly considered the presence of a continued and directed movement of electrons as the most essential sign of the living state.

Oxidation-reduction reactions like polymerization can easily be reproduced in coacervate drops in model experiments. Specifically, we (38) set up a system which the following diagram illustrates:

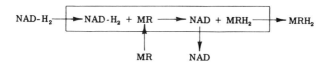

Reduced nicontinamide-adenine dinucleotide (NAD-H$_2$) entered the drop from the external medium and transmitted its hydrogen to dye

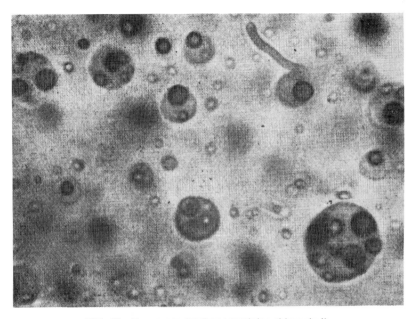

FIG. 20. Coacervate droplets containing chlorophyll.

adsorbed by the drop (MR); the latter was reduced and given off into the external medium. Free energy is released during the process. We accelerate it by adding to the drop an oxidoreductase (NAD-H$_2$-dehydrogenase), isolated by Gel'man (39) from the surface membrance of bacteria. However, this process can also occur spontaneously, and proceeds considerably more rapidly in the drop than in the surrounding medium because the reacting substances are spatially closer to each other (shortening the distance over which the electrons must flow).

The energy released in the electron (hydrogen) transfer is mainly dispersed in the form of heat. It cannot be used directly for the synthesis of high energy compounds or the formation of polymers. However, on combining the oxidation-reduction reactions with phosphorylation (which can take place under anaerobic conditions) energy of the pyrophosphates and other high energy bonds accumulates, energy which is readily available for the synthesis of polymers (40). Our preliminary experiments on simulating a coupled anaerobic phosphorylation in coacervate drops gave positive results.

The flow of substances through the drop is accomplished in a simple oxidation-reduction model wholly by the energy of the reacting substances themselves, releasing, in the process, a transfer of electrons (hydrogen). However, exogenous energy such as the energy of the light falling on the drop may also be used for this purpose. The source of such energy in the initial period of the genesis of life could have been the short-wave ultraviolet light which deeply penetrated the pre-actualistic atmosphere. Subsequently, a significant role came also to be played by less energy-rich quanta of long wavelength light which had to be absorbed by specific substances (pigments) in the drops. This phenomenon was reproduced by Serebrovskaya and Evstigneev (41) in a model experiment in which chlorophyll was included in the coacervate drop (Fig. 20).

If this sort of drop is placed in a solution of ascorbic acid and methylene red and illuminated with visible light, an oxidation-reduction photochemical process occurs which can be represented by the following diagram:

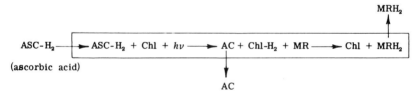

Ascorbic acid entering the drop gives its hydrogen to chlorophyll; this process occurs only with the mediation of light quanta. The chlorophyll then transfers H_2 to a dye entering from the external medium, and then itself is regenerated to its initial state. The reaction products Ac and $MR-H_2$ pass into the external medium.

In this experiment oxidation-reduction reactions took place in the drop many tens of times more rapidly than in the surrounding solution.

A similar type of photochemical reaction could take place in coacervate drops of the "primitive soup," but in the pre-actualistic epoch these reactions were accomplished at the expense of ultraviolet light energy and did not require chlorophyll (or porphyrin derivatives in general). In particular, in this way ATP could be directly formed in coacervate drops in a reaction similar to Ponnamperuma's. The chlorophyll-containing model presented above, however, indicates that longer wavelength radiation could have been used in photochemical reactions at the very beginning of the emergence of the actualistic epoch. We now have before us the task of combining all three types of reactions of primitive metabolism (oxidation-reduction, coupled phosphorylation, and polymerization) in a single coacervate model. The artificial construction of such a model is only a question of time. But it is now clear that such stationary systems should be able not merely to exist for a long time but also to grow in a solution of material which we might justifiably expect to be present, not to say abundant in the "primitive soup" of the Earth.

In addition to this, our simulation of systems indicates that the regular genesis of such a system in the "primitive soup" could be accomplished by natural means. For this it was only necessary to create coacervate drops which were capable of sustaining more rapid reaction rates than in the external medium; at first polymerization, then oxidation-reduction and the phosphorylation linked to it. But such conditions must easily arise; at least, in those drops which are composed of material capable of absorbing very simple organic or inorganic catalysts from the external medium, similar to those used in the model experiments with dyes. By such means the drops constantly acquired an increased concentration of catalysts in comparison with the surrounding medium even when the mass of the drop increased because of its growth by polymerization of monomers of the surrounding solution.

Such systems, which actively interact with the surrounding medium, which have dynamic stability and which are capable not only of maintaining themselves but growing in the "primitive soup," we will for convenience tentatively call "protobionts" in the future.

Protobionts already had a considerably more complex and complete organization than the static coacervate drops but were, however, many times simpler than even the most primitive living things.

Only in a further evolutionary process could these protobionts, arising naturally, be converted into primordial organisms, providing a beginning to life on Earth. We still do not know how to reproduce synthetically this evolution in model experiments, but all the same we can trace with a large degree of accuracy its subsequent progress on the basis of the abundant data which we can obtain by a comparative study of various forms of organization of metabolic processes in the most primitive modern organisms (42).

REFERENCES

1. A. Oparin. *Comp. Biochem. Physiol.* **4**, 377 (1962).
2. R. Goldacre. "Surface Phenomena in Chemistry and Biology," p. 276. Pergamon Press, Oxford, 1958.
3. A. Wilson. *Nature* **183**, 318 (1959); **184**, 99 (1959).
4. S. W. Fox and T. Fukushima. *In* "Problemy evolyutsion. i tekhnich. biokhimii" ("Problems of Evolutionary and Technical Biochemistry"), p. 93. Izd. "Nauka," Moscow, 1964.
5. R. Young and E. Munoz. Cited by Fox (4).
6. S. W. Fox, K. Harada, and J. Kendrick. *Preprints, Intern. Oceanog. Congr.,* p. 80. Am. Assoc. Advance. Sci., Washington, D.C., 1959; S. W. Fox and S. Yuyama. *Ann. N.Y. Acad. Sci.* **108**, 487 (1963).
7. R. Young. *In* "The Origins of Prebiological Systems" (S. W. Fox, ed.). Academic Press, New York, 1965.
8. A. Oparin. *In* "Ocherk dialektiki zhivoi prirody" ("Outline of Dialectics of Living Nature"), Izd. "Mysl," 1965.
9. H. Bungenberg de Yong. *Protoplasma* **15**, 110 (1932).
10. H. Bungenberg de Yong. "La coacervation." Hermann, Paris, 1936.
11. H. Bungenberg de Yong. *Koninkl. Ned. Akad. Wetenschap., Proc.* **50**, 707 (1947).
12. H. Bungenberg de Yong. *Kolloid-Z.* **80**, 221 (1937).
13. H. Booij and H. Bungenberg de Yong. *Protoplasmalogia* **1**, 1962 (1956); P. Koets. *J. Phys. Chem.* **40**, 1191 (1936); D. Dervichian. *J. Sci. Appl.* **2**, 210 (1949).
14. K. Serebrovskaya. Diss. Avtoreferat. (Dissertation. Author's Abstract.). Izd. "Nauka," 1964.
15. C. Perret, *J. Gen. Microbiol.* **22**, 589 (1960).
16. A. Oparin, K. Serebrovskaya, N. Vasil'eva, and T. Balaevskaya. *Dokl. Akad. Nauk SSSR* **154**, 471 (1964).
17. A. Oparin, *In* "The Origins of Prebiological Systems" (S. W. Fox, ed.). Academic Press, New York, 1965.
18. T. Evreinova. Diss. Avtoreferat. (Dissertation, Author's Abstract.). Moscow State University, 1962; T. Evreinova and A. Kuznetsova. *Dokl. Akad. Nauk SSSR* **124**, 688 (1959); T. Evreinova. *In* "Problemy evolyutsion. i tekhnich. biokhimii." ("Problems of Evolutionary and Technical Biochemistry"). Izd. "Nauka," Moscow, 1964.
19. T. Evreinova and A. Kuznetsova. *Biofizika* **6**, 288 (1961); **8**, 395 (1963).
20. A. Oparin, I. Stoyanova, K. Serebrovskaya, and T. Nekrasova. *Dokl. Akad. Nauk SSSR* **150**, 684 (1963).
21. T. Evreinova, *Dokl. Akad. Nauk SSSR* **141**, 1224 (1964).
22. H. Bungenberg de Yong and Mallee. *Kuninkl. Ned. Akad. Wetenschap., Proc.* **56**, 203, (1953); H. Bungenberg de Yong, C. Van Somoren, and F. Klein. *Ibid.* **57**, 1 (1954).
23. K. Serebrovskaya, N. Vasil'eva, and N. Mkrtumova. *Dokl. Akad. Nauk SSSR* **129**, 910 (1964).

24. T. Evreinova, A. Pogosova, T. Tsukanova, and T. Larionova. *Nauchn. Dokl. Vysshei Shkoly* No. 1, 159 (1962).
25. A. Oparin. "Life, its Nature, Origin and Development." Academic Press, New York, 1961.
26. Heraclitus of Ephesus. Fragments. English translation by K. Freeman; Ancilla to the pre-Socratic philosophers, Oxford Univ. Press, London and New York, 1952.
27. R. Schoenheimer. "The Dynamic State of Body Constituents." Harvard Univ. Press, Cambridge, Massachusetts, 1942.
28. I. Prigogine. "An Introduction to the Thermodynamics of Irreversible Processes." Thomas, Springfield, Illinois 1955.
29. A. Pasynskii and G. Dechev. *Izv. Akad. Nauk SSSR, Ser. Biol.* No. 4, 497 (1961); A. Pasynskii. *Usp. Sovrem. Biol.* (1957); A. Pasynskii and V. Slobodskaya. *Dokl. Akad. Nauk SSSR* **153**, 473 (1963).
30. C. Hinshelwood. "The Chemical Kinetics of the Bacterial Cell." Oxford Univ. Press (Clarendon), London and New York, 1947.
31. A. Oparin. *Nova Acta Leopoldina* [N.S.] **26**, No. 165, 87 (1963).
32. A. Oparin. *Proc. Intern. Symp. Enzyme Chem. Tokyo Kyoto, 1957.* Maruzen Co., Tokyo, 1958; A. Oparin. *Advan. Enzymol.* **27**, 347 (1965); B. Libl, N. Khalunka, and I. Molek. *In* "Problemy evolyuts i tekhnich. biokhim." ("Problems of Evolutionary and Technical Biochemistry"). Izd. "Nauka." Moscow, 1964.
33. A. Oparin, T. Evreinova, T. Larionova, and I. Davydova, *Dokl. Akad. Nauk SSSR* **143**, 80 (1962).
34. A. Oparin and K. Serbrovskaya. *Dokl. Akad. Nauk SSSR* **122**, 661 (1958).
35. A. Oparin, K. Serebrovskaya, and T. Auerman. *Biokhimiya* **26**, 499 (1961); A. Oparin, K. Serebrovskaya, S. Pantskhava, and N. Vasil'eva. *ibid.* **28**, 671 (1963).
36. K. Serebrovskaya. *In* "Problemy evolyutsion. i tekhnich. biokhimii" ("Problems of Evolutionary and Technical Biochemistry"), p. 127. Izd. "Nanka," Moscow, 1964.
37. A. Kluyver and C. van Niel. "The microbe's Contribution to Biology." Harvard Univ. Press, Cambridge, Massachusetts, 1956.
38. A. Oparin, K. Serebrovskaya, and S. Pantskhava. *Dokl. Akad. Nauk SSSR* **151**, 235 (1963).
39. N. Gel'man, I. Zhukova, and A. Oparin. *Biokhimiya* **28**, 122 (1963).
40. A. Kotel'nikova. *In* "Problemy evolyutsion. i tekhnich. biokhimii" ("Problems of Evolutionary and Technical Biochemistry"), p. 80. Izd. "Nanka," Moscow, 1964.
41. K. Serebrovskaya, V. Evstigneev, V. Gavrilova, and A. Oparin. *Biofizika* **7**, 34 (1962).
42. A. Oparin. *Proc. 5th Intern. Biochem. Congr., Moscow, 1961* Vol. 3. Pergamon Press, Oxford, 1963.

CHAPTER 5

EVOLUTION OF "PROTOBIONTS" AND THE ORIGIN OF THE FIRST ORGANISMS

Studying the successive stages in the evolution of carbon compounds, we have already examined a fairly high level of organization, rising several steps in the process. We have seen that each preceding step in development served as a basis for emergence of the next. Thus, appearance of hydrocarbons on the Earth's surface was responsible for the formation of the "primitive soup," in which protein- and nucleinlike polymers emerged, in addition to relatively simple organic compounds. The formation of such high molecular weight polymers, although the monomers in the polymeric chains were still not arranged in orderly fashion, led to the segregation from the common solution of the "primitive soup" of individual complex formations like coacervate drops. If even initially these formations were static, they later inevitably must have become dynamic. But the coacervate drops, or other similar individual systems, however, formed in the original hydrosphere of the Earth, and set off from the environment, were immersed not merely in water but in a solution of various salts and organic compounds. These substances penetrated the drop and reacted chemically with themselves and with substances in the drop itself, thus transforming it into an open system, the very existence of which depended on the compatibility of the reactions occurring in it. Without such compatibility, but constantly interacting with the environment, the system would very rapidly have been broken down to disappear as an individual form. And if the interaction between medium and system was broken off for some reason, such a static system would thereby be cut off from the total process of evolution.

Thus, always staying within the bounds of physical and chemical relationships, we have come to the emergence of "protobionts"; their organization to a certain extent can be understood by a study of our models.

127

However, even the most primitive of organisms differ widely from all the systems mentioned above, including the protobionts, by having an exceptionally high "purposiveness" of internal organization, a particular alignment of their metabolism. The countless chemical reactions occurring in living beings not only are strictly coordinated into a single metabolic network, not only are harmoniously combined into a single order of constant self-renewal, but also the whole of this order is unalterably directed to a single goal — the uninterrupted self-preservation and self-reproduction of the entire living system in regular conformity with environmental conditions (1).

We will use the term "purposiveness" here (for lack of a better term) to denote the coordination of the organization of the whole system, its self-preservation and self-reproduction, and in addition to this to denote the adaptability of structure of its separate parts to the most ideal and most harmonious completion of those functions necessary for life, which each part brings to the system as a whole.

The high adaptability of individual organs to the functions which they must perform and the total "purposiveness" of the whole organization are remarkably apparent in even a surface acquaintance with higher living beings. It was noted by man long ago and has found its expression in the entelechy of Aristotle. Its essence seemed mystical and supernatural until Darwin explained rationally and materialistically the origin of this "purposiveness" in the concept of natural selection.

But the "purposiveness" of organization is indigenous not only to higher beings, it permeates the whole living world from the top down to the most elemental forms of life. As we have see above, the correlation between structure and function characteristic even of proteins and nucleic acids is evident even at the molecular level, but only if these molecules are biogenic in nature, are offspring of life. In inorganic nature, away from dependence on life, there is no "purposiveness." Thus we would vainly seek its explanation in the laws of the inorganic world. The emergence of purposeful intracellular organization characteristic of living beings — that organization which is biological metabolism and the dynamic structures of protoplasm related to it — can be understood only on the basis of the same principles which govern the emergence of the "purposiveness" of morphological structure of higher organisms; that is, the interaction of the organism and environment, and the Darwinian principle of natural selection. This new biological law arose in the very process of establishing life and later on occupied a leading place in the development of all living matter.

The organization of protobionts introduced by us in the previous chapter can serve as a starting point for evolution on the way to the establishment of life. The important point in this organization was the fact that the protobionts were not only dynamically stable systems, that they not only could be maintained indefinitely in the "primitive soup," obtaining matter and energy from the surrounding environment, but that when a certain combination of reactions occurred within them they possessed the ability to increase in volume and weight, to grow like the coacervate models which we constructed. During this growth, the protobionts preserved unchanged to a certain degree the form of organization inherent in them. In particular, with the synthesis of more and more new polymeric molecules, a certain role in preserving the constancy of composition of the growing protobionts could have been played by the replication of the polynucleotides contained in them, although this replication based on complementarity was at this stage of evolution still very imperfect. However, the main thing was that the protobionts maintained the constancy of rate relationships and the compatibility of the reactions carried out in them (2). As the protobionts expanded, the original increased concentration was always maintained of the simpler inorganic or organic catalysts which were selectively absorbed from the external medium, just as were the dyes in our model experiments.

It is not very probable that protobionts could always grow as a single mass in the Earth's primary hydrosphere. Under the influence of external mechanical forces (for example, the shock of waves or surf) they must have been divided up, just as are drops of emulsion when they are shaken. Thus, the larger the size attained by any individual protobiont during its growth, the greater the odds for its being broken up into smaller daughter formations. To a certain degree, the latter must have continued the same interchange with the environment which was inherent in the original protobiont, since they were merely small fragments, the parts of a formation relatively homogeneous throughout its entire mass.

Of course this effect could not be compared in any measure in its continuity and accuracy with the self-reproduction of even the most primitive of existing organisms. When the protobionts expand and divide, all sorts of aberrations and changes could very easily occur, particularly if conditions of the environment change. However, all this taken together inevitably led to the original "prebiological natural selection" which brought about the subsequent evolution of protobionts on the pathway to the formation of the first living matter.

In the recent scientific literature, a number of opinions have been expressed regarding the justification of the use of the term "natural selection" as being applicable only to living matter. According to widely held opinions of biologists (3), natural selection is a specifically biological law, and cannot be applied to nonliving objects, particularly to our protobionts.

However, it is wrong to think that at first life emerged, and then biological laws; or on the contrary, that first biological laws were established and then life. This statement of the question is reminiscent of the old scholastic argument as to which came first, the chicken or the egg.

Dialectics oblige us to regard the evolution of living things and the formation of biological laws as an indissoluble unity. This is why it is entirely permissible to consider that protobionts — those systems which were the starting point for the genesis of life — underwent evolution not only as a result of what were properly physical and chemical laws, but also of embryonic biological laws, among them prebiological natural selection. Here we can draw a parallel in the formation of mankind, that is, with the emergence of a social form of the movement of matter even higher than life, which, as is known, was not so much the result of biological factors as of factors which molded society, particularly the work activity of our ancestors; these social factors emerged at a very early state of homogenesis and then were later perfected. Thus as the emergence of mankind is not the result of biological laws alone, so the emergence of living things must not be reduced to an effect of the laws of inorganic nature (4).

To illustrate this, let us cite the following specific example, which to a certain degree is capable of experimental verification. Let us suppose that in a certain solution two types of systems are present (like our models). One has a coordinated combination of reactions leading in certain environmental conditions to the synthesis and expansion of the system as a whole. In the other system, on the contrary, this coordination is disturbed and the processes occurring here lead to a relatively slow synthesis or even to a predominant decomposition. It is apparent that this second type of system will constantly lag in growth and will eventually disappear, being superseded by the first type which is more adaptable to the given conditions.

Only in this elementary sense is it necessary to recognize "prebiological natural selection." To begin with, it is expressed in the fact that every disturbance of the harmony of reactions necessarily causes the death and the disappearance of this so-called "unsuccessful" individual system.

On the contrary, if some protobiont were to show an increase in the rate of oxidation, coupling, and polymerization, an improvement in its coordination or, in general, any change in metabolism favoring more rapid synthesis and growth either in changed or unchanged external circumstances, then this system would naturally take precedence over others, and larger and larger numbers of specimens would begin to form. On this basis there should be a gradual improvement in organization in the overwhelming mass of growing and multiplying protobionts.

This would primarily affect their catalytic apparatus as the most important factor in the organization of metabolism based on the coordination of the rates of individual reactions which make up the metabolism. At the stage with which we are dealing there could not, of course, be any question of such complex substances with specific intramolecular structures as the enzymes of contemporary organisms. The only catalysts available for the protobionts would be very simple organic or inorganic compounds present in considerable amounts in the "primitive soup." Salts of iron, copper, and other heavy metals, for example, could significantly speed up electron (hydrogen) transfer reactions. It is true that their catalytic action is incomparably weaker than the action of such enzymes as peroxidase or phenol oxidase. They are, so to say, very "poor" in comparison with enzymic catalysts; however, as Langenbeck (5) has shown, their catalytic activity can be increased considerably by combining them with certain radicals or molecules.

For example, electron (hydrogen) transfer reactions can be accelerated by inorganic ferric ions, if only very slightly. Acceleration is increased somewhat by combining iron with pyrrole. If the iron is incorporated into a tetrapyrrole compound — into a porphyrin ring, the hemin obtained in this manner will have a catalytic activity exceeding by 1000 times that of inorganic iron. As Langenbeck has shown, even such a simple organic compound as methylamine can accelerate the decarboxylation of keto acids just as does the enzyme carboxylase in living cells. However, by itself methylamine is such a weak catalyst that its effect can be determined only at an increased temperature (under autoclave conditions). Incorporation in the methylamine molecule of a carboxyl group (glycine synthesis) increases its catalytic activity almost 20 times. It is increased several times more by incorporation of an aromatic or heterocyclic ring. Proceeding in this way, consciously incorporating into the original molecule more and more atomic groups, Langenbeck obtained his famous "artificial enzyme models," specifically several compounds which exceeded by many thousand times the catalytic activity of the original methylamine.

This way of successive improvement of simple catalysts could also have been used in the evolutionary development of protobionts by natural selection. Calvin (6) has rightly pointed out the possibility of this sort of evolutionary formation of biological catalysts. If they had been simply in solution, however, such catalysts could not have been subject to natural selection, since their ability to catalyze oxidative reactions did not give any advantage to the molecules themselves as far as longevity or increase in numbers were concerned in comparison with other molecules not having this ability. This situation was different for catalysts incorporated within the protobiont system as a whole. Separate parts of complex, catalytically active molecules dissolved in the surrounding medium could be almost completely lacking this activity. However, selectively adsorbed by the protobionts, they were combined into a catalytically active complex, and if this complex accelerated polymerization or other reactions mentioned above (in comparison with reactions in the environment), it established the continuous flowing character of the system, its dynamic stability, and its ability to grow (as we saw in the example of the model experiments).

From this it is clear that the more ideal a certain complex was — that is, the more its molecular structure corresponded to its catalytic function and the more this function was in conformity with other reactions occurring in the given protobiont, the better it was maintained in the given environmental conditions, the more rapidly it grew and multiplied, and as a consequence it occupied a leading place in the progressive evolution of prebiological systems. One may imagine an enormous number of atomic groupings — radicals, organic and inorganic compounds and their complexes — which in some measure possess the ability to catalyze reactions necessary for the existence of protobionts, for example, the electron (hydrogen) transfer reaction.

In the numerous protobionts first formed, this transfer function, of course, must have been accomplished by a great variety of catalytic mechanisms with widely varying structures, which not only depended on the composition of the medium from which material for the construction of these chemical mechanisms was extracted, but also depended on the individual characteristics of each protobiont.

However, as a result of the fact that natural selection constantly eliminated the less perfect mechanisms, and hence the systems which contained them, this diversity gradually narrowed. From the boundless sea of chemical possibilities which the "primitive soup" possessed, only a few of the most effective combinations of molecular groups were selected. Thus, the earlier in the development of protobionts such a standardization occurred of a catalytic mechanism, the more universal it must have been for the whole subsequent living world.

As one of the most striking examples of this we must cite NAD (nicontinamide adenine dinucleotide), which participates as a universal hydrogen carrier in numerous oxidation-reduction processes of living cells (7). We find it in all contemporary living beings absolutely without exception, in microbes and in higher plants or animals, in heterotrophs and in autotrophs, in organisms fermenting and oxidizing various sugars, and in living forms having as sources of carbon nutrition phenols and other closely related hydrocarbon derivatives.

This indicates that NAD was selected by living nature from the multitude of similar compounds at a relatively early evolutionary stage of organic matter, at the very source of life. Of course, the abiogenic synthesis in the "primitive soup" of these same adenine derivatives, a primary possibility which we have mentioned before, contributed to this. Adenine derivatives must therefore have entered the developing protobiont from the external environment when the protobiont was first formed. Soon they acquired more and more significance as hydrogen carriers, as they became more complex and better adapted to this process.

Comparative biochemical study has indicated that somewhat analogous groups of compounds are the flavin derivatives (8), those quite universal catalysts of oxidation-reduction processes in modern organisms, then later coenzyme A (CoA) and several other similar compounds.

The data presented permit us to postulate that at a certain stage in evolution of the protobionts, coenzymes acted as accelerators; they replaced simpler but less perfect organic and inorganic catalysts.

Coenzymes now play a prominent role in the metabolism of all modern organisms. They are few in number, but any one of them is an extremely widely distributed universal accelerator of biological processes, which is indicated by their very early development in the genesis and development of life (9).

It was entirely unnecessary in the first stages of evolution that coenzymes be completely synthesized in the protobionts themselves. The uniformity of concentration of catalysts necessary for the dynamic stability of the developing protobionts could be maintained simply by the entry of these relatively complex compounds or components forming them from the environment. In this connection, it may be recalled that many contemporary, even highly organized living things in spite of the unconditional requirements of their metabolism, cannot synthesize certain of the coenzymes and must obtain them from the environment in the form of vitamins (10).

However, the formation of processes leading to the synthesis of coenzymes, from simpler and simpler compounds is a significant step forward on the way to progressive evolution of protobionts. It created conditions possible for the existence of protobionts in a less complex medium, but in place of this required a coordinated interaction of a large number of reactions necessary for synthesis.

Horowitz (11) on the basis of his investigations on the fungus *Neurospora* gave an interesting outline of this type of evolving complexity of synthesizing ability. To some degree it can also be used in the case which we are developing of the synthesis of coenzymes and evolving protobionts. The nucleus of this scheme is as follows: Let us assume that some primitive organism or, generally, some sort of open system requires for its metabolism the more or less complex organic compound A. If this compound is present in a prepared form in the surrounding medium, the system can take it up directly, even without any special chemical adaptations for its synthesis. But if a deficit of substance A arises in the environment, or if it more or less entirely disappears, only those systems can continue to exist in which some way is found to form new chemical mechanisms permitting the synthesis of compound A from a simpler substance B, C, or D, present in sufficient quantities in the surrounding solution. Later on this same process must also be repeated for substance B, when it will have been exhausted into the surrounding medium, etc.

Thus the original forms of the evolving systems must have been completely heterotrophic, that is, on the whole they had to depend on the complex composition of the surrounding medium. Further evolution proceeded toward the formation of more complex and multilinked processes in the systems, decreasing the dependence of the system on the environment.

To a certain degree this evolutionary pathway can be demonstrated even in the example of our coacervate models. Drops having merely polymerizing ability can maintain themselves and grow only in a medium containing high energy phosphorus compounds. The inclusion of oxidation-reduction reactions and conjugated mechanisms permitted the drops themselves to synthesize high energy compounds and made possible the growth of the drop in a less complex and special medium. Similarly, the existence of protobionts using for acceleration and coordination of these three reactions only ready-made catalysts occurring in the medium surrounding them, was very limited by these conditions, and the catalysts themselves would have to be comparatively primitive, only slightly specialized substances. Thus, the significant step forward is the synthesis in the protobiont itself of more effective catalysts (coenzymes) from less complex components of the medium. But for this to take place, an additional new synthetic reaction would be required, linked to the three reactions named above. Thus in the process of growth and natural selection of protobionts a system of metabolic reactions inside the protobionts had to be perfected and also made more complex.

Purely theoretical considerations stated by several contemporary authors (12) compel us to suppose that the progressive evolution of such open systems as our protobionts has proceeded not only by way of an improvement in the coordination of the small number of reactions mentioned above, but also in the direction of an increasing number of these reactions, a lengthening of the chains formed by them, branching of these chains, and their closing off into constantly repeating cycles. This actually is confirmed by the presence in all modern organisms of quite complex metabolic networks consisting of a very large number of reactions. In various representatives of the living world this network can vary widely, as we will see below, but it must contain these three types of reactions which we spoke of above as its individual links. These reactions were already inherent in protobionts, and composed the basis of their rudimentary metabolism, but in the process of evolution they were added to many times and became more complicated by the inclusion of newer reactions and combinations of reactions.

However, the longer and more diverse the chain of reactions became, the larger the number which entered the overall metabolic network, the more strictly the rates of individual reactions had to be coordinated; more perfect catalytic mechanisms were necessary in order to attain this. Thus the earlier few and relatively weakly specialized catalysts – coenzymes – were insufficient to resolve this complex problem, and the next very important step in the progressive evolution of protobionts was the creation of a whole arsenal of new powerful catalyst-enzymes; that is, proteins with an intramolecular structure extremely well adapted to the fulfillment of their catalytic functions.

Dixon and Webb (13) in the conclusion of their fundamental book rightly noted that the problem of the genesis of enzymes is one of the most difficult questions to solve. They consider completely possible the abiogenic polymerization of amino acids or their activated derivatives in the primordial aqueous solution of organic substances on the Earth.

It could have been and must have been here that numerous high molecular weight polypeptides, and various proteinlike substances were formed. However, the initial emergence of such an enzyme pathway, according to Dixon and Webb, is completely improbable.

Enzymes possess exceptional activity and specificity of catalytic action simply because their intramolecular structure is very perfectly adapted to the accomplishment of their biological function; but this adaptability, of course, cannot arise accidentally in a simple solution of organic compounds.

It is true, we are already acquainted with the formation of a certain adaptability of catalytic apparatus to the functions carried out by it in primitive metabolism from the example of coenzyme formations. First, however, these formations, as we have seen, cannot simply be accomplished in the solution of the "primary soup," but in the dynamically stable, growing and multiplying protobionts, on the basis of the natural selection of these individual systems. Second, it is necessary to clarify the difference in principle between coenzymes and modern protein enzymes.

Many enzymes (proteids) contain in their catalytically active center a particular prosthetic coenzyme group. But in the proteids, as well as in other enzymes (proteins), the side chains of amino acid residues combine in their active centers; the side chain containing hydroxyl, amino, carboxyl, sulfhydryl, imidazole, indole, and other similar groups of well-known chemical composition. In some cases these groups, like the coenzymes, can become catalysts but their catalytic activity is very ordinary and they cannot compare in any way with the enzymes themselves, which indeed have an enormous ability to accelerate a particular reaction (14). Thus, for example, the enzyme hexokinase speeds up the interaction of ATP with glucose more than 10^{11} times, and alcohol dehydrogenase, the oxidation of alcohol more than a billion (10^9) times, etc. The very high specificity of enzymic activity is also exceedingly remarkable, which is the basis of that aphorism which E. Fisher uttered in his day, that any enzyme fitted its substrate as a key does a lock.

The circumstance that each specific enzyme catalyzes only a strictly specific reaction among the multitude of other potentially possible reactions for each substrate has tremendous significance for the organization of metabolism, not only from the point of view of a general acceleration of the whole living process, but also in relation to the definite sequence of metabolic reactions and its overall direction.

The following elementary scheme can illustrate this. Let us assume that some organic substance A can be converted into the substances B, C, D, etc. In the diagram the rates of these conversions are represented by radius vectors, the length of which characterizes the rate of the corresponding reaction.

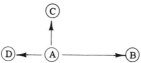

We see from this diagram that the rate of the reaction $A \rightarrow B$ is 7 times greater than the rate of the conversion $A \rightarrow D$, and this is ½ as fast as the reaction $A \rightarrow C$. It is apparent that after a certain time has elapsed, when the whole supply of substance A disappears, there will be present in the mixture obtained 70% B, 20% C, and 10% D. Most of substance A was converted to B, that is, this conversion followed the path of the fastest reaction.

If, in this case, we provide some sort of effect which uniformly increases the rates of all the possible reactions (for example, an increase in temperature), the ratio of the yields of the final products is not changed in any way. But if we introduce into the initial mixture a catalyst specifically accelerating only the reaction A → D many times, for example, a million times, and not affecting the rate of reactions A → B and A → C, an entirely different effect will be obtained. Substance ·A is thus almost completely converted to D, and B and C are reduced to barely perceptible or even imperceptible traces.

Substance D, which has thus arisen, like any other organic compound has many chemical potentialities but in the presence of a new specific enzyme, it will only follow the one path laid out for it by that enzyme. Thus there arises a chain of consecutive reactions coordinated with one another in time.

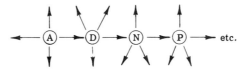

But such a chain is possible only when there is a whole series of catalysts and each is specific only for a corresponding link of the chain. Yet in a large degree the significance of this specificity emerges only when the chain branches. Then we must turn our attention not only to the necessary presence of all catalysts (enzymes) specific for each link of the chain, but also to the correlation of their catalytic activity. Any intermediate compound at the branching point of the chain can actually follow not one, but at least two different routes, each route being governed by its specific enzymes and the relative magnitude of its catalytic activity. When the number of links in the metabolic chain was small, the order of metabolism could be regulated by such relatively imperfect catalysts as coenzymes. The increasing complexity of the metabolic system was possible only on the emergence of very powerful and stictly specific catalysts – the enzymes.

The coenzymes themselves could be formed purely kinetically as a result of the coordination of the rates of a relatively small number of reactions. On the contrary, the synthesis of enzymes required enormously more complex mechanisms, these being not only the formation of molecular groups composing the active center, but also a strictly regulated construction of the whole protein molecule (or at least a signifi-

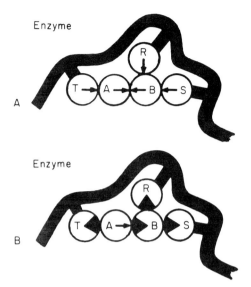

FIG. 21. Schematic diagram of a hypothetical reaction occurring on the surface of an enzyme in which two substrates, A and B, and three catalytic groups of enzymes, R, S, and T participate. The orientation resulting in the maximum reactive capability is shown by the arrows in A. The blackened sectors designate that part of the surface of the molecule on which the reaction can take place (B). The active groups are kept in their places by special structures of the enzyme, the substrates by the specificity of adsorption processes.

cant part of it) both in respect to an accurate sequence of amino acid residues in the polypeptide chain and the spatial packing of this chain into the protein globule. Although many modern data are evidence of the fact that catalytic activity and specificity of enzyme action are chiefly determined by the active center, which is only a very small part of the protein molecule, we nevertheless know that the residual struc- ture of the polypeptide chain is important. We must regard it as a sort of skeleton, which has as its purpose the creation of a corresponding arrangement in space of groups which determine catalytic activity and specificity.

Koshland (15) considers that the high catalytic quality of enzymes is due to the reciprocal orientation of substrates and catalyst, to a signifi- cant degree and perhaps entirely. In explanation he gives a simple scheme – a "triple complex" between two interacting substrates and an enzyme with three catalytically active groups arranged at different parts of the polypeptide chain (Fig. 21). In addition to this, Koshland considers that we must not regard the enzyme as a simple "template" or "mold" for the substrate; that the configuration of the enzyme undergoes significant changes in the course of reacting with the substrate. As a model he turns to a comparison with a glove. "The glove," he writes, "is not a three-dimensional negative of a hand. It may be in any number of shapes, but only after introduction of the hand is the close three-dimensional fit obtained. In the enzyme case a substrate is presumed to induce a proper alignment of catalytic groups so that enzyme action ensues." This concept of the "flexible" enzyme requires a very complete ordering of the structure of the enzymic protein.

In contrast to this, the arrangement of amino acid residues in the first proteinlike polymers could only have been random and nonordered. These polymers could serve adequately as material for the formation of coacervate drops and enter into the composition of developing protobionts, but they either entirely lacked catalytic activity or were very poor catalysts.

Certainly, as the drops or protobionts became more dynamically stable, polymerization of the amino acids entering them from the external environment could have formed combinations of amino acid residues which to some degree were inherently favorable for catalytic activity. However, this formation of an improved polypeptide created an advantage for the embryonic protobiont only when the polypeptide synthesis occurring on further growth of the protobionts invariably led to combinations of amino acid residues catalytically profitable to the protobiont. During the usual "nonordered" polymerization of amino acids entering the protobiont from outside, this advantage was quickly lost, leveled out in the growing protobiont.

Thus the creation of some type of organization was very important for further progressive evolution of protobionts — that organization which would permit the synthesis of polypeptides already having some more or less stable arrangement of amino acid residues and which would provide constancy of intramolecular structure in the newly synthesized polypeptides. Such an organization could not be based wholly on kinetic considerations, on the stability of the sequence of reactions in the metabolic chain; it required fundamentally new spatial ("matrix") mechanisms. An exceptionally important role in this respect fell to the lot of the polynucleotides.

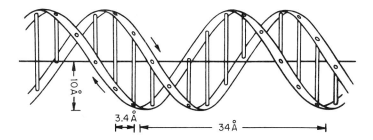

FIG. 22. Structural model of the DNA molecule (after Watson and Crick).

As was indicated above, only the "nonordered" synthesis of polynucleotides was possible in the primary "nutrient soup," polynucleotides
having a random arrangement of mononucleotide residues in their molecular chains. But as soon as this chain reached a certain size even the
"nonordered" partially polymerized polynucleotides reacted with polypeptides and other polymers of the "primitive soup" and together with
them were segregated from the general solution in the form of coacervate drops, as we have demonstrated experimentally.

The polynucleotides differ, however, from other polymers in the
following remarkable feature. More than 10 years ago Watson and Crick
(16), on the basis of a series of steric and chemical considerations,
showed that two polynucleotide chains can combine with each other in
a double helix only complementarily (Fig. 22); that is, so that a
particular purine or pyrimidine base of one chain is attached by hydrogen bonds only to some different base not identical with the first, of
another chain (for example, adenine combines with thymine or uridine,
and guanine always with cytosine) (Fig. 23). This situation was confirmed experimentally, first on simple synthetic polynucleotides consisting entirely of identical mononucleotide residues. Thus, for example, it
was shown that polythymidylic acid complemented a polynucleotide
chain containing no other base but adenine (polyadenylic acid). Later,
the significance of the complementarity of polynucleotide chains was
demonstrated in the synthesis both of natural nucleic acids (DNA), and
artificial polynucleotides made up of all four bases, when the succession
of bases in one such chain determined their sequence in another (17).

According to Schram (18), this effect also occurs in the abiogenic,
nonenzymic synthesis of polynucleotides. Thus, for example, in his
experiments on the polycondensation of uridine monophosphate, the

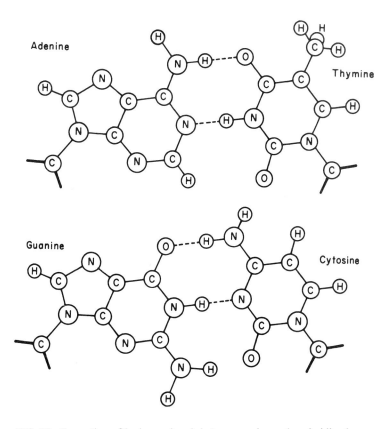

FIG. 23. Formation of hydrogen bonds between purine and pyrimidine bases.

synthesis of uridylic acid went many times more rapidly in the presence of polyadenylic acid which was complementary to it, and on the contrary, the formation of uridylic acid accelerated the synthesis of the polyadenylic chain.

In the simple aqueous solution of the "primitive soup" this sort of accelerated formation of one certain polymer would only lead to its continual accumulation in the form of unusual organic deposits. Even in this case, if during these syntheses some sort of accidental change occurred, for example, the inclusion in the primarily homogeneous chain of other mononucleotides, these so-called "mutations" could not be effective in the further evolution of organic matter. The circumstance that one or several residues of adenine in the polyadenylic chain were replaced by another base would not give this chain any sort of advantage in comparison with the previously unaltered chain in its further "multiplication." This is why no selection of specific polynucleotide molecules could occur in a simple aqueous solution. An entirely different situation arose when these molecules were incorporated into an entire system, containing other polymers, particularly proteinlike polypeptides, such as took place when the coacervate drops formed in the "primitive soup."

In this case a particular sequence of mononucleotides in the polynucleotide chain could influence the combination of amino acid active groups in the polypeptides forming into a system and this combination could exert its effect on the organization of the primitive metabolism of the whole system. If such a way of altering metabolism was favorable in the sense of the preservation and growth of the entire system, it was maintained by natural selection and was consolidated in the developing system by replication of the polynucleotides being formed. In the opposite case they were kept out of the subsequent evolution of protobionts, as a result of the death of the systems harboring these changes. Thus, properly, not the polynucleotide chain itself, but the entire system in which it had brought about certain changes in metabolism, was subject to selection.

Owing to the successes which biochemists in recent years have attained (19), we know that the synthesis of proteins in modern organisms with a specific sequence of amino acid residues in their polypeptide chains, and in particular the synthesis of enzymes, is accomplished by extremely complex and complete mechanisms, preexisting in any cell (20). We give in Fig. 24 a simplified outline of this.

This mechanism includes about 20 different specific enzymes — amino acid activators, approximately 20 different specific types of low molecular weight soluble ribonucleic acid (sRNA) (the so-called transport RNA), and a certain type of matrix determining the sequence in which amino acids are strung along the protein chain. The large molecules of information mRNA functioning in specific subcellular particles, the ribosomes, apparently act as these matrices (Fig. 25). The molecules of mRNA themselves are synthesized in the cellular nuclei under the control of DNA. The latter fulfills two functions: first, it provides for

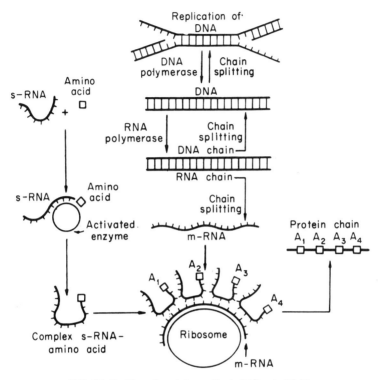

FIG. 24. Outline of protein synthesis (After A. Rich).

the reproduction of itself by replication during cell division, stabilizing in the molecules being formed the conformation of mononucleotides inherent in the original molecular chain; second, DNA has the ability to construct complementary chains of mRNA corresponding to its structure and through it to specify the conformation of amino acids in protein synthesis (21). In addition to the substances and chemical mechanisms cited, probably a specific assemblage of enzymes is still required for the synthesis of peptide bonds, for the removal of prepared molecules of protein from the ribosomal matrix, and also for the reactivation of sRNA. In addition to this, in the respective stages of synthesis in the system, energy in the form of adenosine triphosphoric acid (ATP) synthesized in the mitochondria would have to be supplied.

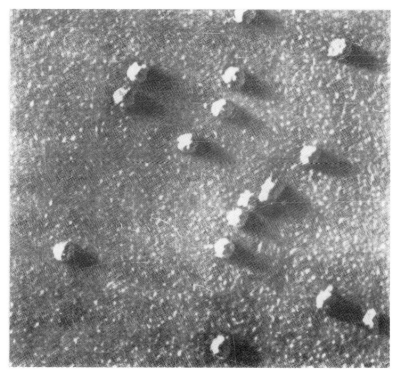

FIG. 25. Ribosomes from bacterial cells (X 30,000).

The process of protein synthesis, probably, takes place in the following manner: at first, amino acid, activated by reaction with ATP forms aminoacyladenylate. In order to carry out this stage, the presence of an enzyme is necessary, and apparently different amino acids require different enzymes. Aminoacyladenylate, which evidently remains fastened to the activating enzyme, then is transferred to sRNA. Then each particle composed of a complex of sRNA and activated amino acid is bound to a specific site on the matrix mRNA in the ribosome. Thus, the amino acids are collected onto the polypeptide chain in a certain order, depending on the arrangement of mononucleotide residues in the polynucleotide chain of mRNA. In essence then, a particular amino acid (or more accurately its complex with sRNA) corresponds to a particular mononucleotide triplet in the mRNA chain; thus, for example, aspartic acid-guanine uracil, adenine (GUA); alanine-cytosine, cytosine, guanine (CCG); phenylalanine – uracil, uracil, uracil (UUU) etc. (22).

It is entirely apparent that this type of exceptionally complex and complete protein synthesizing apparatus of contemporary organisms could have been formed only as a result of a very long evolution of preexisting systems; the stability of these systems was basically dynamic in character and depended on the compatibility of their metabolic reactions and interaction with the environment. The formation of a spatial "molecular matrix" apparatus is an additional superstructure, raising this older form of organization to a new unprecedented height of perfection of the conformation of chemical processes and accuracy of system self-reproduction, but preserving in principle the previous dynamic stability of living systems formed in this way.

Commoner (23) rightly warns of too great an enthusiasm for the matrix concept of self-reproducing DNA, pointing out that in respect to biological systems it inevitably leads to preformation. Even in modern organisms, the specifics of protein synthesis do not depend on the spatial organization of macromolecules and on matrix duplication alone. An important role in this process is also played by the kinetic properties of the reacting systems, and not just the structure of the original molecules. This is all the more justifiable for systems preceding the appearance of life. Thus in the formation of strictly ordered polynucleotide chains, spatial matrices for protein synthesis were by no means the starting point of prebiological evolution, but, on the contrary, a final higher stage.

A very interesting sketch of such evolution was given recently by Rich (24) in his review paper devoted to this question. However, Rich thinks of this evolution occurring entirely on the molecular level, as a process of gradually increasing complexity and orderly arrangement of the intramolecular structure of polynucleotides and polypeptides by natural selection merely in solution outside of any sort of complete system.

In contrast to this, it has seemed to us that first, the interactions of even randomly structured polypeptides and polynucleotides in the "primitive soup" must necessarily have led mainly to the formation of multimolecular individual systems (coacervates and protobionts); and second, that individual molecular structures were not subjected to natural selection, but only those individual dynamic systems which had best solved the problem of coordinating their primitive metabolism with the preservation and growth of the systems in given environmental conditions.

As a result of all that we have said, we should be able (although still very hypothetically, of course), to map out some successive stages in the prebiological evolution of the protein synthesizing apparatus in the first stage of this evolution. Dissolved in the "primitive soup," nucleotide monomers were assembled purely abiogenically into randomly built chains of polynucleotides independently of any catalytic action of proteinlike substances. Rich, using the results of Kornberg's experiments on the synthesis of AT-copolymers in the absence of primers, and data on the isolation of this type of polymer from some organisms (25), expresses the very interesting thought that the primary polynucleotides could have been composed of fewer different units than modern nucleic acids. Thus, for example, the first polynucleotides could have contained only two complementary bases — adenine and thymine, which are comparatively easily obtained abiogenically. Similarly, the randomly structured polypeptides which were synthesized abiogenically in the "primitive soup" at the same time as the polynucleotides, but entirely independently of them, would also have had fewer different units. This, in particular, could have been caused by the fact that not all 20 amino acids of modern protein molecules were necessarily formed abiogenically under the influence of ultraviolet light or atmospheric electric discharge.

However, as experiments presented above from our laboratory have shown, molecules even of very primitive, simply and randomly built amino acid and mononucleotide polymers synthesized at the same time in a common solution and having attained a definite size, are joined without fail into multimolecular clusters and are separated from the solution in the form of coacervate drops. The formation of these systems greatly enhanced the acceleration of polymerization, but also within the drop this polymerization of primitive peptides and nucleotides could have proceeded entirely independently of each other. Coupled energetic reactions arising in the drops and protobionts at first contributed to and then totally determined the polymerizations. The rate of all reactions occurring in the system depended on the entry into it of primitive catalysts and coenzymes, while amino acid polymers could play only a minor and accidental role. The role of the phenomenon of replication of primitive polynucleotides was also minor.

In the second stage, for the first time, the interaction of molecules of various polymers, gradually increasing in complexity, became significant. Even in contemporary organisms, the synthesis of polypeptides and specialized proteins with a comparatively very simple chemical structure can be found outside of the ribosomal matrix structure, but nucleoside phosphates are required for the reaction.

Thus, for example, the following outline of polypeptide synthesis has been established in some microorganisms:

$$\text{nucleoside triphosphate} + \text{amino acid} \rightarrow$$

$$\text{nucleoside-diphosphate} + \text{peptide} + \text{orthophosphate};$$

furthermore, for each of the four nucleotide triphosphates (ATP, GTP, CTP, and UTP) a certain group of amino acids exists which it can activate (26). Thus the structure of a polypeptide to a certain degree is determined by the nature of the nucleoside triphosphate participating in the reaction.

If, in a system similar to our protobionts, participating in polymer formation, the nucleotide were to contain ribose, 3^1 and 5^1-hydroxyl groups will be used in polymerization; however, the third group located in the 2^1 position remains free, and it can join to itself the amino acid corresponding to a given nucleotide through an ether linkage. "At this stage," writes Rich, "We theorize this would be the beginning of a prototype system in which the polymerization of a nucleic acid molecule is coupled with the assembly of a series of amino acids. The amino acids might then be subsequently polymerized once they are organized in a linear assembly by their attachment to the adjoining nucleotides." Thus it is readily apparent how the sequence of monomers in the polynucleotide can to a certain degree specify some order of arrangement of amino acid residues in the synthesized polypeptide.

However, it is at this point that basic difficulties begin for hypotheses involving all phenomena on the molecular level. In this case it is necessary, in order for the polypeptides thus synthesized to affect in their turn the synthesis of polynucleotides, that there be a specific catalyst for this process. Only in this way could it have a definite advantage from the point of view of prebiological selection. In contemporary organisms this actually takes place, but in the simple solution of "primitive soup" such conformity would be extremely improbable, a rare "lucky accident." An entirely different situation is created if we will examine all the phenomena in the entire dynamic system, in the growing and multiplying protobiont. In this case the combination of amino acid residues being formed by the influence of polynucleotide cannot affect directly the polymerization of nucleotides, but it catalyzed certain reactions occurring in a particular protobiont just as did the coenzymes which enter it or are synthesized in it. If the acceleration of this reaction in combination with other processes occurring in the protobiont were favorable for its existence, the protobiont would obtain an advantage over other similar systems in the rate of its growth and multiplication.

In the opposite case, if the emerging amino acid sequence lacked catalytic activity favorable for the protobiont, or this activity decreased the dynamic stability of the system, the given amino acid combination would disappear under the influence of natural selection together with the embryonic system.

Thus it was not these polynucleotides free to replicate, and not even the polypeptides emerging under their influence and having a certain sequence of amino acid residues which themselves were subject to selection but entire systems, protobionts, with a primitive but more or less well-developed metabolism, coordinated or not coordinated with certain conditions for existence. The role of the polynucleotides in this case was confined to spatially consolidating the stability of synthesis of catalytically advantageous amino acid combinations in the growing and multiplying protobionts; they served as a stabilizing factor in protobiont evolution.

On this basis, a gradual increase in complexity and refinement both of polynucleotides and polypeptides occurred simultaneously — a lengthening of their chains, an increase in the diversity of the links incorporated into these chains and the establishment of a definite strict sequence of these links in the polymeric chains.

Molecules of proteinlike polymers, as well as the polynucleotides controlling their synthesis became better regulated, better adapted to those functions which they fulfill in evolving prebiological and biological systems. This is presumed to be the beginning of the emergence of enzymes, the adaptability of which to their catalytic functions has been continuously improved during the process of evolution. Moreover, the polynucleotides were also undergoing evolution; their complex functions were later broadly differentiated in the process, which was the third and final stage in the perfection of the protein-synthesizing apparatus.

In contemporary organisms we have two types of nucleic acids — DNA and RNA, the functions of which differ widely, in spite of their great chemical similarity of their molecules. DNA was specialized functionally in the molecular replication cycle. Because there is no hydroxyl group at the second carbon atom in the deoxyribose, DNA is not capable of binding amino acids to itself and thus does not participate directly in protein synthesis. Although it is metabolically comparatively very inert, it provides for a relatively high stability of self reproduction of itself, and for the transfer of the genetic information contained in its polynucleotide chain by means of the replication of ribonucleic acid. The function of the latter is directly involved in protein synthesis, in the establishment of a strict and definite conformation of amino acid residues in protein molecules.

However, we must note that this functional specialization of RNA is not absolute. In principle RNA can also carry genetic information; this occurs, for example, in RNA-containing viruses.

Thus there is a basis for postulating that the initial forms of nucleic acids were RNA-like polymers, capable of both storing and transferring hereditary information, as well as organizing amino acid sequences for protein synthesis. However, the division of these functions between the two nucleic acids, of which the less metabolically active specialized in self-reproduction, while the other participated directly in protein synthesis, was an extremely progressive step in the process of evolution and thus was consolidated by natural selection (27).

It is clear that contemporary protein-synthesizing systems and the enzymic proteins formed as a result of their action are just the final result of a complex competition between many different systems, which began at a very early stage of the genesis of life. In particular, it is necessary to visualize that in the evolution of protobionts, between the original slightly catalytic action of the polypeptides and the exceptionally efficient modern enzymes, there were tested and rejected at least as many, and possibly more, organizational variants than, for example, between the fins of the shark and the human arm. The overwhelming mass of catalytic variants formed during evolution was eliminated by natural selection. We now extract from contemporary organisms only very efficient enzymes, although with more intensive study of the question we even now can observe some evolution of these catalysts (28).

Still in large measure this concerns the evolutionary development of a combination of individual enzymic reactions into a common metabolic organization and the gradual formation of a spatial macrostructure in modern living beings, requiring various degrees of evolutionary development. The fundamental principles of this organization in space and in time were still founded on the process of gradual development of the protobionts, and thus they are common to all living systems.

However, later on, evolutionary development of primitive organisms emerging on this basis continued by different routes, and we can judge this development, and consequently, its sources, on the basis of the tremendous factual material which modern comparative biochemistry has now accumulated.

REFERENCES

1. A. Oparin. "Life, its Nature, Origin and Development." Academic Press, New York, 1961.
2. R. Eakin. *Proc. Natl. Acad. Sci. U.S.* **49**, 300 (1963).
3. P. Mora. *Proc. Conf. Origin Prebiol. Systems, Wakulla Springs, Florida, 1963.* Academic Press, New York, 1965; P. Gavaudan. Introduction to the French edition of A. Oparin, "L'origine de la vie sur la terre." Masson, Paris, 1965.
4. A. Oparin. "Life as a Form of the Motion of Matter." Acad. Sci. USSR, 1963.
5. W. Langenbeck. "Die organischen Katalysatoren und ihre Beziehungen zu den Fermenten," 2nd ed. Springer, Berlin, 1949; *Advan. Enzymol.* **14**, 163 (1953).
6. M. Calvin *Am. Sci.* **44**, 248 (1956); *Science* **130**, 3383 (1959).
7. N. Kaplan, M. Swartz, M. French, and M. Ciotti. *Proc. Natl. Acad. Sci. U.S.* **12**, 481 (1956).
8. G. Mahler. *Proc. 3rd Intern. Congr. Biochem. Brussels, 1955* p. 264. Academic Press, New York, 1956.
9. A. Yurkevich, E. Severin, and A. Braunshtein. *In* "Enzymes," p.147. Izd. "Nauka," Moscow, 1964.
10. "Katalyticheskie funkstii vitaminov" ("Catalytic Function of the Vitamins"). Foreign Language Publ., Moscow, 1953.
11. N. Horowitz. *Proc. Natl. Acad. Sci. U.S.* **31**, 153 (1945); N. Horowitz and S. Miller. "Current Theories on the Origin of Life." Springer, Vienna, 1962.
12. J. Pringle. *New Biol.* **16**, 54 (1954); C. Perret. *J. Gen. Microbiol.* **22**, 589 (1960).
13. M. Dixon and E. Webb. "Enzymes," 2nd ed. Academic Press, New York, 1964.
14. D. Koshland. *J. Cellular Comp. Physiol.* **47**, Suppl. 1, 217 (1956); *Advan. Enzymol.* **22**, 45 (1960).
15. D. Koshland. *In* "Horizons in Biochemistry" (M. Kasha and B. Pullman, eds.). Academic Press, New York, 1962.
16. I. Watson and F. Crick. *Nature* **171**, 737 and 964 (1953).
17. A. Kornberg. *Science* **131**, 1503 (1960); S. Ochoa. "Nobelstifftelsen." Stockholm, 1960.

18. G. Schram, H. Grotsch, and W. Pollman. *Angew. Chem.* **73**, 619 (1961).
19. A. Belozerskii. "Nukleinovye kisloty i ikh biologicheskoe znachenie" ("Nucleic Acids and their Biological Significance"). Izd. "Znanie," 1963.
20. G. Webster. *Ann. Rev. Plant Physiol.* **12**, 113 (1961).
21. F. Crick. *In* "Struktura i funkstii kletki" ("Structure and Function of Cells"), p. 9. Izd. "Mir," 1964; M. Nirenberg. *Ibid.*
22. S. Ochoa. *In* "Horizons in Biochemistry" (M. Kasha and B. Pullman, eds.). Academic Press, New York, 1962.
23. B. Commoner. *In* "Horizons in Biochemistry" (M. Kasha and B. Pullman, eds.). Academic Press, New York, 1962.
24. A. Rich. *In* "Horizons in Biochemistry" (M. Kasha and B. Pullman, eds.). Academic Press, New York, 1962.
25. N. Sueoka. *J. Mol. Biol.* **3**, 31 (1961).
26. A. Bogdanov. *Abstr. Rept. 1st All-Union Biochem. Congr., Leningrad, 1963* Vol. 1, p. 13. Izd. Akad. Nauk SSSR, Moscow, 1964.
27. A. Spirin. *Abstr. Rept. 1st All-Union Biochem. Congr., Leningrad, 1963* Vol. 1, p. 13. Izd. Akad. Nauk SSSR, Moscow, 1964.
28. N. Kaplan. *Proc. 5th Intern. Congr. Biochem., Moscow, 1961* Vol. 3, p. 101. Pergamon Press, Oxford, 1963.

CHAPTER 6

FURTHER EVOLUTION OF THE FIRST ORGANISMS

The chemical history of the emergence of life on Earth might properly conclude with the formation of the first organisms. After this, matter
entered a new biological stage of development; this marked the beginning of the evolution of living beings from very primitive organisms to
modern highly organized plants and animals.

However, intensive study of this purely biological evolution can
furnish us with much that is useful in understanding the emergence of
life itself, its initial establishment. We cannot directly observe this
process in nature, since all of its intermediate links — the very primitive,
imperfect forms of organization of living matter have long since disappeared, wiped from the face of the Earth by natural selection, even,
apparently, without leaving behind them reliable fossil traces. Nevertheless these traces are preserved in the organization of protoplasm in the
living beings being contemporaneous with us.

"Among all natural systems" — write Zuckerkandl and Pauling (1) —
"only living matter is distinguished by the fact that in spite of the very
significant changes it has undergone, it has kept a way of recording in
its organization a tremendous amount of information reflecting its history."

Studying this organization we can obtain some objective data on the
very earliest forms of life on Earth. Investigations of metabolism, in
particular, can give us much in this respect (2), since metabolism is the
regular sequence of biochemical processes which forms the basis for the
organization of all life. A primitive form of metabolism, a certain
alignment of oxidation-reduction, linking and polymerization reactions,
were already inherent in the very first organisms, but in the process of
biological evolution they were constantly adding on new links in metabolism, and new internal chemical mechanisms which these links were
dependent on — catalysts, coenzymes, polynucleotides, enzymes, supermolecular structures, etc.

This made it possible for newly emerging life to put to better use a wider range of sources of matter and energy necessary for life, to be less dependent on the composition of the environment and less subject to the risk of extermination during some alteration in its composition.

Modern comparative biochemical investigations (3) have most definitely shown that the new evolving biochemical reactions and their combinations, far from replacing the previous metabolic links, only supplemented them; forming, as it were, an added "superstructure" on the older metabolic forms.

We can show these forms quite clearly by studying metabolism in present living organisms, striving to trace in the tremendous diversity of modern metabolic systems traits of similarity, traits of organization, which are most common to all living beings without exception (4).

Just as anatomists. studying and comparing the structure of organs of specific animals, reproduce the picture of their evolutionary development and permit us to look back into their remote past, so biochemists by a comparative study of the metabolisms of various organisms can approach the very springs of life, recognizing its most primitive forms of organization.

With this approach, the following two cardinal conclusions are clearly evident: (1) metabolism of all contemporary living beings is based on that form of organization which has adapted to constant use of ready-made organic compounds as the original building material for the biosynthesis of proteins, nucleic acids, and other component parts of protoplasm and as a direct source of energy for these biosyntheses. The overwhelming majority of biological forms now inhabiting our planet can exist only by constant replenishment of prepared organic matter from the external environment. Both higher and lower animals belong here, among them the majority of Protozoa, the great majority of bacteria, and all forms of fungi. This fact alone is extremly significant, but the metabolism of green plants also and other autotrophic organisms which have already worked out for themselves the ability to synthesize organic compounds from CO_2, water and mineral salts are based on the same groups of reactions and the same chemical mechanisms as are the rest of the heterotrophs. Autotrophic mechanisms are only added superstructures to the basic structure of the organisms having them. Under certain conditions they can easily be eliminated from metabolism, and the organisms will not then be destroyed, but merely change over entirely to the use of exogenous organic nutrients (5).

This can be very easily shown in the example of the less well-organized photoautotrophs — algae — both under natural conditions and in the laboratory. Such experiments have long since established that when organic compounds are introduced into sterile cultures of algae they are directly assimilated. The process of assimilation can be developed simultaneously with the process of photosynthesis, but in some cases, photosynthesis can be totally eliminated, and the algae then change over completely to a saprophytic form of life (6).

Many forms of blue-green algae and other algae in nature use polluted water extensively for direct nutrition of organic matter. This fact is obvious from their especially luxuriant development in liquid sewage and other similar places rich in organic matter.

We find heterotrophic bases of nutrition in all higher plants, although the photosynthetic apparatus has reached the peak of its development in them. However, in higher plants only the chlorophyll-bearing cells have it. All the remaining colorless tissue builds its metabolism on the basis of the use of organic substances common to all other living things; the organic compounds enter from the photosynthesizing organs. Leaves, particularly, change over to this sort of metabolism when there is no light.

This position is solid confirmation of our hypothesis stated in the previous chapters about pathways for the genesis of the first living beings. The assimilation of organic matter soluble in the surrounding aqueous medium and its transformation into component cell parts was, of course, an absolute requirement of the initial form of metabolism in all primitive organisms. It was derived from protobionts capable of incorporating within themselves organic matter of the "nutritive soup." Further evolution at first led to selection of those forms of organization which made possible more rapid and improved assimilation of this matter.

(2) All modern organisms base their biochemical systems for obtaining the energy necessary for life from organic matter on the mechanism of anaerobic decomposition, which is amazingly similar in all organisms. We are readily convinced of this when we examine closely the biochemical processes of metabolism both in lower and higher modern organisms.

It is true that the majority of these organisms are now unconditional (obligatory) aerobes, and only a very limited number of modern primitive biological forms can exist for very long in the absence of free oxygen, or else do not require oxygen at all and lead a completely anaerobic existence. This situation is entirely natural in the present epoch. Respiration is an incomparably more efficient energy-creating process than anaerobic metabolism. As soon as free oxygen appeared in the Earth's atmosphere, organisms in the process of evolutionary improvement must have widely adopted the aerobic form of life. But it is quite noteworthy that at the base of the energy metabolism of all modern aerobes lies the same chain of reaction not requiring free O_2, as in primitive anaerobes; oxidation reactions using gaseous oxygen which are specific for aerobes only supplement this anaerobic mechanism, practically universal for all life. In contrast to this anaerobic mechanism, the oxidative "superstructure" is by no means universal. Different groups of contemporary organisms have significantly different systems. Polyphenol oxidase, for example, which is characteristic of higher plants, is lacking in the respiratory mechanisms of animals in which terminal oxidative enzymes are active.

Thus we see that combinations of anaerobic reactions are at the base of the "trunk of the tree of life" while the respiratory mechanisms joined it later when this "trunk" had divided into separate "branches," into separate groups of more highly organized living beings, and each of these groups had solved the problem of using the free oxygen generated at this period of the Earth's existence in ways specific for each (7).

These evolutionary-biochemical data are perhaps one of the most conclusive proofs of the fact that life arose when there was still a reducing atmosphere and hydrosphere, since if we were to assume that these events occurred in the presence of free oxygen, the existing relationship between anaerobic and aerobic mechanisms of metabolism in present-day organisms would be completely incomprehensible. In exactly the same way the data presented above on the heterotrophic basis of nutrition convinces us that the organic matter in the environment surrounding the organisms, formed abiogenically or deposited on the Earth, was the original source of energy and structural materials for the first living beings.

Now, the evolutionary-biochemical investigations of present living organisms to a certain degree permit us to formulate principles of organization of the first living beings, long ago gone from the face of the Earth, which of course had a very much more primitive metabolism than any of the modern microbes, and in this respect closely approached our hypothetical protobiont. As the protobionts, the first organisms were heterotrophs and anaerobes but they built their catalytic apparatus on the basis of a sequential ordering of intramolecular structures of polypeptides and polynucleotides able to replicate; it was owing to this that the catalytically favorable arrangement of amino acid residues was fixed in a position for further protein synthesis during the growth and multiplication of the organisms. The collection of enzymes constantly being improved on this basis permitted the primary organisms having them to join to their metabolic chain a greater number of links, to create a longer chain of reactions which meshed together well and which made considerably more efficient use of material from the environment and favored more rapid growth and multiplication.

In contemporary heterotrophic organisms which have traveled a long road of development we find that form of organization of constructive metabolism in which very simple low molecular weight compounds such as ammonia, acetic acid, glycine, succinic acid, keto acids, etc. act as starting materials for the synthesis of all the complex ingredients of protoplasm (8). These compounds originate as fragments of organic sources of nutrition entering from the environment and decomposed in destructive metabolism. Thus in modern metabolism the constructive and destructive divisions are very intimately connected and are only two aspects of a single process.

But for this organization of metabolism, a very complex, multilinked system of reactions is necessary, in which a tremendous number of separate chemical acts very accurately and continuously are coordinated with time. The progressive evolution of the original organisms moved forward only with the development and improvement of these multilinked systems.

The combination of oxidation-reduction, linking and polymerization reactions inherent in the original organisms were constantly being supplemented and made more complicated by new reactions and their combinations during later evolution. In modern organisms, therefore, we find the original simple reactions only as component elements of the overall metabolic system, only as separate, although very significant links in long reaction chains and cycles.

For the overwhelming majority of modern organisms the main source of carbon nutrition and the basis of energy metabolism are some of the carbohydrates which enter the organism from the environment or are synthesized by it. It is difficult to say whether this form of metabolism was at the base of the still unbranched trunk of the "tree of life" or whether it was a very early and main branch.

Some doubt has been cast on the primacy of the monopoly of carbohydrate nutrition, specifically by investigation of metabolism in several microorganisms isolated from the soil of oil-bearing regions (9). These living beings are not capable of assimilating sugar, and as a sole source of carbon and energy they use the hydrocarbons of oil and their related derivatives such as paraffins, phenols, toluol, salicylic acid, etc. It is too bad that there are not irreproachable data which would permit us to say with confidence whether these organisms are shoots from the "central trunk of the tree of life" directly preceding us, or whether their unique metabolism was secondary in origin. And in this and the other case it is very significant that the main hydrogen carriers of the whole living world in oxidation-reduction reactions and the "coupled" energetic mechanisms are present in both types of organisms (10).

But be that as it may, it may be considered that the main branch of the "tree of life," almost monopolizing development during the further evolution of life, was the carbohydrate branch of energy metabolism. It is evident that at the very base of this branch, a very complex combination of reactions is formed, a multilinked chain of chemical transformations, inherent in all present living organisms, even the most primitive, that is to say, those found at the most extreme depths of biological evolution, which only our modern comparative biochemical methods can plumb. This is illustrated in Fig. 26 by the chain of reactions transforming glucose to pyruvate. It consists of 10 basic reactions, the rates of which are so coordinated that they follow each other in a strictly determined sequence.

The foremost reactions in the chain are the ones already well known to us from our model experiments with coacervates – the transfer of hydrogen and the "accumulation" of energy, mediated by NAD and ATP. But in distinction from our models and the protobionts, these basic links of the chain, even in the most primitive organisms, are supplemented by new reactions, related first of all to the very important preparation of the substrate and its oxidation-reduction transformation. This preparation, on the one hand, leads to the elevation of the sugar molecule to a higher energy level, which is attained by phosphorylation by means of ATP, and on the other hand, to the decomposition of particles of hexose diphosphate obtained in this way into two molecules

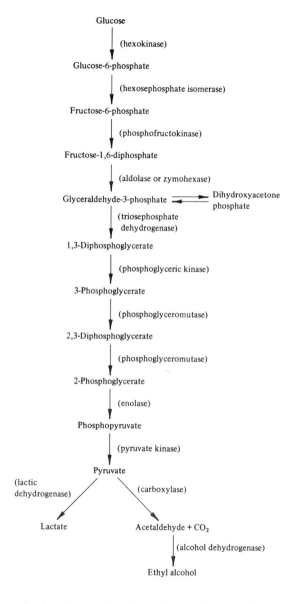

FIG. 26. Outline of alcohol and lactic acid fermentation.

of triose phosphate. This sort of complicating mechanism in the chain is a significant step forward on the route of rationalization of energy metabolism. It is interesting to note that this very old sequence of transformations, the basis of biological metabolism, does not require any sort of supermolecular spatial organization, any sort of protoplasm structure, and can be realized simply in a homogeneous solution, the prototype of which is the pressed yeast juice of Buchner.

In this way the organization of metabolism here is based only on the coordination of individual reactions in the chain with time, although the reaction rates necessary for this accurate combination require the participation of such modern and specific catalysts as only enzymes can be.

This chain of reactions leading to the conversion of glucose to pyruvate apparently possessed very great biological efficiency, because, formed at comparatively early stages of the existence of life (perhaps only 2 billion years ago) it has continued throughout the whole development of living matter and has been preserved not only in primitive microbes but also in the metabolism of higher plants and animals.

In these two branches of living nature so far removed from each other we can observe merely a forking of the route in further transformation of pyruvate. In plants pyruvate can be broken down into CO_2 and alcohol (as in alcohol fermentation) and in animals it is reduced to lactate (as in lactic acid fermentation).

These two types of anaerobic decomposition of carbohydrates are the most widespread, and among many modern microorganisms they are the high road in the evolutionary development of destructive metabolism. But of course in the past other variants must have arisen, many of which were completely rejected, destroyed by natural selection, while others have been kept up to the present time as thin twigs on the main branches of the "tree of life." And actually, we now find a large number of various types of anaerobic fermentation, but at the same time each of these types is characteristic of a very limited group of lower organisms. The difference begins with pyruvate, however; pyruvate is always formed by a single pathway universal for all life (Fig. 27).

Examination of this outline scheme shows that, depending on the type of fermentation, besides alcohol, CO_2 and lactic acid, other different fermentation products can be formed, including formic, acetic, propionic, succinic, butyric and other higher fatty acids; ethyl, propyl, and butyl alcohols, acetyl methyl carbinol, acetone, gaseous hydrogen, etc. It

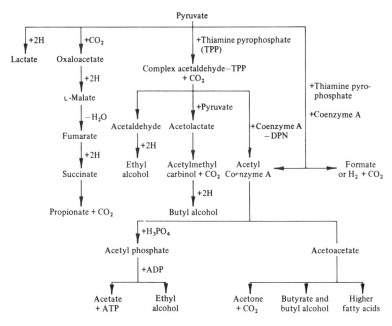

FIG. 27. The principal reactions of pyruvate in various types of bacterial fermentation.

is characteristic that this diversity is attained by the inclusion of a very small number of "supplementary" reactions and by the use of almost always the same catalytic mechanisms. The whole business is reduced to a variation of combinations of almost always the same individual chemical acts.

Of course, both alcoholic and lactic acid fermentations as well as the other anaerobic fermentations we have mentioned must not be regarded as some sort of isolated destructive processes. All of them are closely related to the synthetic reactions of constructive metabolism, supplying energy and building materials for them. For example, pyruvic acid, occupying a key position in all types of fermentation, can easily be converted into one of the most important amino acids – alanine – with the participation of ammonia and hydrogen carried by pyridine nucleotide according to the following equation:

$$\text{pyruvate} + \text{ammonia} + \text{NAD-H}_2 = \text{alanine} + \text{NAD} + \text{H}_2\text{O}$$

The same is true for other keto acids also. Fragments of molecules formed as intermediate fermentation products can be condensed into long open chains, into closed aromatic and heterocyclic molecules, serving as links for the formation of biologically important polymers.

In this way, the better coordinated in time the individual reactions are for a given type of metabolism, the less energy is dissipated, the greater the percent of required nutrients goes into the construction of the system, and as a final result, the reactions prove to be improved biologically and more promising in an evolutionary respect.

However, the highly interrelated metabolic pathways we have presented were not of course the only possible reaction systems in the reducing conditions of the atmosphere and hydrosphere in the period we are examining of the development of life on Earth. Alternative routes for anaerobic transformations of organic substances can be found in modern organisms. Krebs and Kornberg (11) consider that one of these ways is the pentose phosphate cycle, which is given in outline in Fig. 28, as derived from their book.

In contrast to the forms of fermentation described above, in this process the molecule of phosphorylated sugar does not break down into two molecules of triose phosphate, but right after the formation of glucose 6-phosphate it is subjected to anaerobic oxidation, and CO_2 is split off from the phosphogluconic acid obtained in this way, as a result of which a phosphorylated derivative of pentose is formed. In this way, ribose, which is very important in the construction of nucleotides and nucleic acids, is formed.

The pentose phosphate cycle is a much less universal form of carbohydrate metabolism than alcohol or lactic acid metabolism. This cycle is the main pathway of carbohydrate splitting in only a few microorganisms; in the majority it only supplements the usual forms of fermentation. As experiments with labeled glucose have shown, only a small part of the sugar is split by way of the pentose phosphate cycle and apparently its main purpose is the formation of ribose.

We can imagine that the pentose phosphate cycle arose some time after the basic types of fermentation had formed. "It is very probable" — write Krebs and Kornberg in this connection "that the oldest organisms did not require it. These organisms could use pentoses which were abundant in the medium surrounding them. It was only when the pentose content decreased significantly, that the pentose phosphate

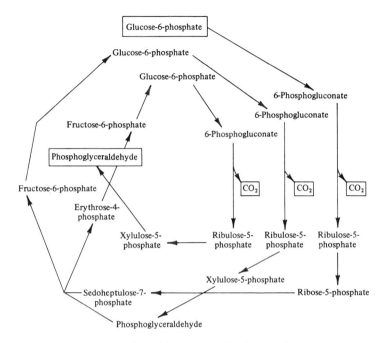

FIG. 28. Outline of the pentose phosphate cycle.

cycle became part of the metabolism of organisms." Its later origin is also indicated by the fact that the cycle included the basic fermentation reaction in addition to some newly emerging reactions. Furthermore, for future accounts it is very important to note that reactions occur in the pentose phosphate cycle which are of outstanding significance in photosynthetic processes.

The numerous forms of metabolism which we have discussed are characteristic of the diversity of the ways by which the evolution of organisms proceeded, still without free oxygen in the environment, but in the presence of ready-made organic matter, that is, in anaerobic and heterotrophic conditions.

Although the supply of organic compounds contained in the "primitive nutritive soup" was replenished by geological endogenic processes and by the meteorites and comets falling on the Earth, nevertheless it must have diminished as life developed. This greatly aggravated the struggle for existence and was a new powerful factor in the future evolution of organisms. During their development, those metabolic systems began to arise which permitted not only a more efficient assimilation of organic substances of the "primary soup," but also used other more simple forms of carbon nutrition and sources of free energy more widely distributed in the external medium.

Early processes related to the generation and further use of energy included mainly dehydrogenation and condensation reactions catalyzed at first by primitive catalysts and then also by enzymes. The primordial organisms utilized the surrounding medium, rich in organic compounds with reducing properties, for coupling dehydrogenation with the release and utilization of energy. As a consequence of this, the surrounding medium became gradually more oxidized. It could only be reduced with the expenditure of energy coming from outside. The inexhaustible source of this energy was and now is solar radiation.

On the still lifeless Earth with its pre-actualistic reducing atmosphere, organic matter was photochemically transformed mainly by energy from short-wave ultraviolet light carrying a large store of energy in each of its quanta. This same source of energy could easily have been utilized by primitive organisms during their initial existence. But as a consequence of the gradual enrichment of the atmosphere with even small amounts of free oxygen, an ozone screen was formed, almost entirely blocking the approach of short-wave radiation to the Earth's surface. Thus further development of life proceeded by utilizing the longer wave radiation, since it remained available for organisms in the transitional epoch of the Earth's existence.

However, the utilization of quanta of visible light less rich in energy required the creation of a quite complex internal organization which could emerge only gradually, only at a comparatively high level of the development of life.

It is known that the energy of visible light can be used for the accomplishment of a series of oxidation-reduction processes in the presence of organic dyes capable of absorbing this light. According to Terenin (12) the dye molecule absorbing light acquires a high reactive capability permitting it to receive or donate electrons (hydrogen) and thus to accomplish even those oxidation-reduction processes which in themselves could not be completed in the dark without the addition of light energy. Porphyrins, in particular, can act as such dyes in organisms.

Porphyrins, their molecular structure based on a ring of four pyrrole

units, apparently appeared during the evolution of organic matter at a
somewhat later stage than adenine or flavin derivatives, for example. As
the investigations of Krasnovskii (13) and Szutka (14) have shown, free
oxygen is necessary for the synthesis of porphyrins. Thus they could
have arisen only in the transitional period when the Earth's atmosphere
began to be enriched with molecules of O_2.

It is interesting to compare this case with the fact that the overwhelming majority of modern anaerobes lack porphyrins, and the hydrogen transport necessary to them they accomplish by means of flavin
enzymes. In contrast to this, porphyrins play a decisive role in the
respiratory process of aerobes, and are basic links in their electron
transport chain (15). The iron porphyrins are particularly significant in
this respect; the commonly known representative of this group is the red
blood pigment — hemin. The biological function of the iron porphyrins
(electron transfer during the oxidation-reduction reactions) is successfully carried out in the dark also, in the absence of light (16). Consequently, that important property of porphyrins related to their color,
the ability to absorb light, is not even utilized here.

However, in contrast to the iron porphyrins, the nonmetal containing
porphyrins and in particular, their magnesium complexes, not having the
properties of the usual dark catalysts, are capable of photocatalytic
activity. The work of Krasnovskii (17) and his colleagues has shown that
the magnesium porphyrin complexes, bacterial chlorophyll and chlorophyll of higher plants, as well as the nonmetallic porphyrins (for example hematoporphyrin) acquire the capacity for reduction (taking up an
electron or hydrogen), but absorbing the corresponding quantum of
light, so that this photocatalytic transfer of an electron or hydrogen in
the absence of dark catalytic processes leads to an increase in the energy
level of the photoreaction products; in other words "placing in reserve"
part of the absorbed light energy in a very mobile easily utilizable form
(18).

In the beginning period of the existence of life, when the primordial
organic compounds were abundant in the external medium, visible light
as an energy source could not have been decisively significant for
organisms. But as the ready-made exogenic organic compounds became
exhausted, as their deficit grew in the surrounding solution, greater and
greater advantage in the struggle for survival was obtained by those
organisms which were in a condition to use their porphyrins as photo-

catalysts. They were then able to utilize visible light as an additional source of energy. This first of all allowed the primordial pigmented organism, without having recourse to a significant rearrangement of its entire previous organization, radically to increase the efficiency of its heterotrophic metabolism by using exogenic organic matter much more economically.

The common heterotrophs were forced to convert a considerable percentage of the organic matter which they obtained from the external medium into less utilizable by-products: alcohol, organic acids, etc.

On the contrary, the primordial pigmented organisms used for this purpose "free-of-charge" energy of visible light which released them from the inefficient consumption of exogenic organic matter. This was the initial concept of photochemical reactions, rather than the primary synthesis of organic matter (19).

An example of this may be seen in the study of metabolism in modern pigmented bacteria, specifically Athiorhodaceae (20). From the outside, from the point of view of total balance, the metabolism of these bacteria is characteristic of common heterotrophic metabolism. They grow very well anaerobically in the light, in solutions containing the required organic nutrients (for example, butyric acid or other corresponding compounds). As the bacterial biomass grows, the quantity of exogenic organic substances in the surrounding medium correspondingly decreases, and in addition to this the bacteria give off into the external atmosphere a small amount of carbon dioxide gas.

However, their internal biochemical mechanisms are essentially complex. Like some other organisms, they are capable of fixing atmospheric CO_2. But utilizing the increased energy from the absorbed light of the pigments, the bacteria Athiorhodaceae carry out the photocatalytic transfer of hydrogen, reducing carbon dioxide and oxidizing exogenic organic matter. Thus nonutilizable waste products are not formed, as is necessarily the case in other heterotrophs. Athiorhodaceae in the light use exogenic organic matter almost completely (90% and more) in order to build up their biomass, while in the usual (dark) heterotrophs, nonutilizable waste products absorb the lion's share of nutritive substances (21).

Other pigmented bacteria metabolize according to the same scheme as Athiorhodaceae, but their source (donor) of hydrogen for the reduction of carbon doxide is not organic matter, but hydrogen sulfide. This was shown by the very interesting investigations of van Niel on purple and green sulfur bacteria (Thiorhodaceae) which populate shallow sea inlets and lagoons well-illuminated by the sun and rich in hydrogen sulfide (22).

All these primitive pigmented organisms have the sort of mechanisms which enable them to transport hydrogen or electrons photochemically at the expense of absorbed light energy. But they all must use only the most available substances as an original source of hydrogen, for example, organic compounds, hydrogen sulfide, molecular hydrogen, etc. (23).

The process of progressive evolution of photosynthesis was directed toward the creation of those mechanisms which would permit the utilization as hydrogen donors of a much wider range of substances.

This route of development inevitably led as a final result to the incorporation in the photosynthetic reaction of the most "difficult" but also most "ubiquitous" hydrogen donor – water. As a result, the oxygen in water was released in the molecular form (24).

Some of the contemporary organisms are interesting because they have still preserved in their metabolism traces of a more primitive organization of photosynthetic processes, but they already have the ability to release molecular oxygen from water. They are, as it were, intermediate links between primordial photosynthetics and highly organized photoautotrophs. Specifically, one of these organisms is the green alga *Scenedesmus;* Gaffron (25) has studied its metabolism in detail from this point of view.

However, there is no doubt that the high road in the development of autotrophs was photosynthesis in that form which we now observe in higher plants. The utilization of water as a hydrogen donor by photosynthetic organisms is an enormous step forward on the route of development of biochemical systems, linking the photo stage of the process with reaction cycles leading to the stepwise reduction of carbon dioxide and the formation of molecular oxygen.

But for this to take place, a long evolution of already quite highly developed organisms was required, organisms having a large arsenal of diverse metabolic mechanisms (26). What we know of the photosynthetic apparatus of contemporary plants assures us that this must be so. This apparatus is exceptionally complex and in spite of numerous investigations, still remains far from being completely deciphered (27).

In order to attain the greatest clarity in presentation we will permit ourselves the following comparison, as always, of course, very tentative. Let us take as an example some sort of complex system, performing certain work — an automobile motor. The work of the motor depends not only on its main part, the cylinder block, but also on a series of auxiliary mechanisms, of which some represent a whole collection of parts with specific assignments — preparation and delivery of the fuel mixture, the production of a high voltage current for igniting this mixture, cooling, greasing, transmission of motion, change in speed, etc.

For trouble free operation of the motor, not only the proper functioning of each of these aggregates is important, but also their perfect coordination both in space and in time — synchrony: the spark from the spark plug must flash at a strictly determined position of the piston in the cylinder, the fuel mixture must enter at a certain moment, etc.

Similarly to this, in the photosynthetic apparatus of plants we have not a single type of chemical reaction chain, but a series of cycles of biochemical reactions, entire aggregates of catalytic and photochemical systems. And only when they are perfectly coordinated, when they are constantly interacting, can a reliable effect be obtained. This is attained not only on the basis of a strictly regulated combination of individual reactions with time, but also their location in space, the presence of a definite structure of the photosynthetic apparatus.

In contrast to the situation which we have in the chain of anaerobic fermentation (alcohol fermentation, for example), photosynthesis cannot be produced simply in homogeneous conditions. It requires a definite spatial organization. This requirement became particularly important in connection with the use of water as a hydrogen donor and with the accompanying release of molecular oxygen. In this process only the separate disposition of the initially unstable products of photosynthesis among heterogeneous structures can prevent the thermodynamically more probable reverse course of the reaction. In pigmented bacteria, which carry out photosynthesis under anaerobic conditions without the formation of free oxygen, the pigments are diffused throughout the protoplasm or concentrated at the surface membrane. In contrast to this, the higher form of photosynthesis became possible only when a special complex structural apparatus (the plastid) was formed during plant evolution. Electron microscopic investigation indicates that the plastids consist of colorless stroma in which the "dark" reactions occur and that they also contain the chlorophyll granules which have as their main structural element a protein-lipid membrane. The granules look like stacks of coins. They consist of protein platelets, joined by a chloro-phyll-containing lipid layer (Fig. 29).

Only in this protein-lipid aggregate can the primary photochemical act of splitting water occur. But in order to accomplish the entire photosynthetic process, this act must be harmoniously combined with a long series of other processes, based on enzymic mechanisms or even whole systems and aggregates of them. Using mainly the data of Calvin (28), we give here a very simplified outline (Fig. 30) of the operation of these aggregates each of which can be characterized by its function in the overall process of photosynthesis: (1) formation of molecular oxy-gen, (2) dark fixation of carbon dioxide, (3) reduction of carbon dioxide to the level of carbohydrates, (4) synthesis of sugars from phosphotrioses, (5) formation of "active hydrogen" in the form of the reduced pyridine nucleotide ($NAD-H_2$)*, (6) formation of high energy bonds (ATP).

Light, falling on the chlorophyll-bearing layer of the protein-lipid aggregate, dislodges electrons from it, which go toward the reduction of pyridine nucleotide (in system 5), while the positive charge acts on water, causing it to be oxidized (in system 1); the intermediate products of this reaction are hydrogen peroxide or organic peroxides, which are broken down to form molecular oxygen. Oxygen is mainly given off into the atmosphere, but some of it is utilized in system 6 for oxidative phosphorylation processes.

On the other side (in system 2) dark fixation of CO_2 takes place, which mainly is accomplished by the same routes and by the same mechanisms (CoA for example) that are present in other living beings. As a result of the processes completed in system 2, phosphoglyceric acid is obtained, which is further converted in systems 3 and 4 into various sugars. In system 3 it is reduced to triose phosphate (glyceric aldehyde). But first, "active hydrogen" is necessary for this, which is obtained here in the form of reduced pyridine nucleotide ($NAD-H_2$) from system 5, and second, an easily available source of energy is required (ATP) produced by system 6.

Further conversion of the triose phosphates is accomplished in system 4. It consists either in simple condensation (by which hexose diphos-

*According to the new terminology $NADP-H_2$ is an analog of $NAD-H_2$, includ-ing not 2 but 3 particles of phosphoric acid.

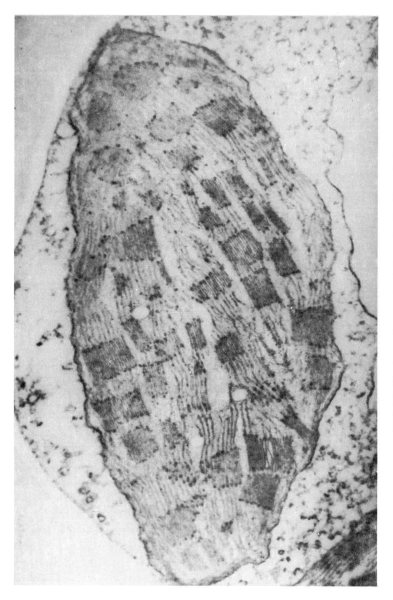

FIG. 29. Electron microscopic photograph of a chloroplast.

phate and then glucose are formed) or in more complex transformations leading to the emergence of phosphoric esters of all the monosaccharides characteristic of the plant kingdom with 4, 5, 6, 7, and 10 carbon atoms. Specifically the phosphoric esters of pentoses are formed here — the ribuloses, which are characteristic intermediate products, as we have seen above, of the pentose phosphate fermentation cycle. Ribulose monophosphate plays a very important role in the photosynthetic process, since it enters system 4 after additional phosphorylation with ATP, where it serves as a primary acceptor of carbon dioxide during dark fixation.

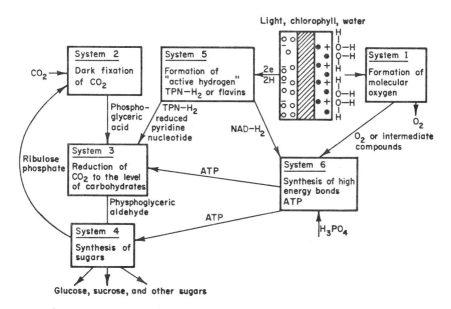

FIG. 30. Outline of the interaction of various systems of compounds in the process of photosynthesis.

Detailed knowledge of the photosynthetic apparatus in green plants indicates that all of its catalytic mechanisms and even entire aggregates of them are not something new in principle. In the majority of cases we find the same or similar mechanisms in various nonpigmented organisms or in photosynthesizing bacteria.

Thus, before the appearance of green plants, before the emergence of modern forms of photosynthesis, these chemical mechanisms existed, but they were not joined into a single complex system. It was precisely this joining of earlier existing "mechanisms" which occurred at the creation of the photosynthetic apparatus. It could be formed during the evolution of organisms only on the basis of previously existing systems and reactions.

The emergence of photosynthesis was an extremely important stage in the process of evolution of the organic world of our planet. It radically changed all the relationships existing up to that time, caused the transition from the pre-actualistic epoch of the Earth's existence to the actualistic, when free oxygen was rapidly accumulated in the atmosphere.

For modern obligatory anaerobes, oxygen is a poisonous gas, strongly inhibiting growth and retarding development. The reasons for this are still not completely clear; apparently they can be different in different anaerobes. The majority of them are able to absorb atmospheric oxygen, which probably is dependent on the fact that flavins — those basic carriers of hydrogen in anaerobic oxidoreductions are autooxidizing agents, compounds capable of being directly oxidized by molecular oxygen (29). In the absence of free oxygen this has no significance for anaerobic metabolism. When oxygen is present this metabolism is disturbed, perhaps, as a result of the breakdown of flavin mechanisms or the formation of hydrogen peroxide. In aerobes hydrogen peroxide is broken down by the iron porphyrins, which the primitive anaerobes lack.

Be that as it may, during the period when free oxygen first appeared and was gradually increasing in the environment, the struggle must have begun for the maintenance of anaerobic life, and in this struggle, those organisms survived which in some way adapted to the change setting in. This was achieved not only by direct withdrawal from immediate contact with the atmosphere, but also by a gradual alteration in the character of metabolism, different in different representatives of the living world. In particular, according to the opinion of MacElroy (30), it was precisely on the basis of the struggle for anaerobic living conditions that bioluminescence first made its appearance, since this was the most effective way of removing oxygen from the sphere of anaerobic metabolism.

In this transitional epoch metabolism formed in that unique group of organisms which are the chemoautotrophs. Just on the edge between reducing and oxidizing conditions, there were the widest theoretical possibilities for the oxidation by molecular oxygen of the reduced inorganic compounds of the Earth's crust.

In the period we have discussed, when free oxygen was beginning to form, these oxidation reactions must have taken place literally at any point on the Earth's surface, since oxidizable substrates were present everywhere. However, abiogenically these reactions happened relatively very slowly and the energy released as a result was lost, dissipated in the form of heat.

Where there was an acute deficit of exogenic organic compounds, those organisms which during their evolutionary development could incorporate these oxidation reactions of inorganic matter into their metabolism, and could form within themselves catalytic mechanisms accelerating these processes and mobilizing their energy for biosyntheses, of course, acquired a great advantage in the struggle for existence and thus were preserved by natural selection and subsequently widely evolved.

At the present, as a rule, we find organisms capable of chemoautotrophic life in nature precisely where deeply buried reduced matter, coming to the surface, has met the molecular oxygen of the atmosphere (31).

Thus, chemoautotrophs at the present time are very important in various cycles of substances. In nature, practically all oxidative processes in reduced nitrogen and sulfur compounds, as well as hydrogen, methane, and partly iron, are associated with the vital activities of corresponding microorganisms (32). The large systematic variegation of groups of chemoautotrophs and the proximity of certain of their representatives to various more primitive heterotrophs to which many are related by transitional organisms, compels us to realize that chemoautotrophs did not just arise once, but that the beginning of their most luxurious development was at that time when there was more diversity of organic forms (33).

The specific conditions of the period we are discussing contributed to this development; first, there was an insufficiency of organic nutrition and a large supply of inorganic sources of energy. However, when the Earth's surface changed to an oxidized state, this supply was quite rapidly consumed, and its replenishment from deeply lying layers of the Earth's envelope occurred comparatively slowly. On the contrary, the balance of organic matter in the biosphere became more and more positive as a consequence of the emergence and rapid development of photoautotrophs

This permitted the basic evolutionary current to return to the old channel of further development of organisms adapted to the nutrition of organic matter. The period of acute insufficiency of this organic matter was a thing of the past, and the only biological memory of this was preserved in a small group of autotrophic organisms capable of chemosynthesis, these being side channels of the main evolutionary current. At present, the main channels of this current are the green plant photoautotrophs and nonpigmented organisms, retaining their former earlier method of heterotrophic nutrition. But after the emergence of photosynthesis, evolution in these organisms also, utilizing prepared organic compounds in their vital processes, was carried out on a completely different biochemical basis. The decisive factor in this respect was atmospheric oxygen, which permitted considerably more efficiency and greater intensity of mobilization of the energy of organic matter. It is apparent that this increased efficiency was based on the same anaerobic mechanisms which earlier formed the foundation of energy metabolism in the early heterotrophs.

But during evolution under the new aerobic conditions, only those organisms were preserved by natural selection and guaranteed future development in which supplementary enzyme complexes and reaction systems arose, enabling them to receive from exogenic organic matter much more energy than previously by complete oxidation of these substances by atmospheric oxygen (34).

The solution of this problem required the formation of two new systems: first, a system for mobilizing the hydrogen which went to waste in anaerobic conditions, secreted by the organisms in the form of unutilized reduced organic compounds (acids, alcohols, etc.) or even in the form of gaseous hydrogen; and second, a system for activating oxygen in order to oxidize hydrogen to water, to accomplish the reaction of oxyhydrogen gas.

The individual mechanisms of the first system are very old. They were basically inherent in anaerobic organisms. We are already well acquainted with NAD, ATP, CoA, etc.; now, with the emergence of aerobiosis, their activity was spread into a series of new products, not participating in the chain of alcohol or lactic acid fermentation. This first chain of carbohydrate transformation was itself preserved, unchanged in aerobes, but in certain places new chains and cycles of reactions were joined to the original; the separate links of the new cycles yielded their hydrogen to NAD or other similar acceptors (for example, flavin derivatives).

Such a junction point of new cycles is plainly evident even in quite primitive facultative anaerobes. First of all, there is pyruvic acid, that key point from which the paths of various forms of anaerobic fermentation branch off in different directions (35). Thus, for example, in *Streotococcus faecalis* pyruvic acid is converted to acetic acid on oxidative decarboxylation, and in contrast to this, in bacteria from the propionic acid fermentation, it joins CO_2 to itself and forms oxaloacetic acid. In higher organisms capable of respiration both these processes are performed on pyruvic acid in the common way. But the matter does not end here, for a multilinked closed chain of transformations arises, called the Krebs cycle (36), or the di- and tricarboxylic acid cycle (Fig. 31). For simplicity, the anaerobic pathway of glucose and pyruvic acid transformation common to all organisms is omitted from the outline. We will not dwell in detail on all links of this complex cycle; many of the mechanisms known to us take part in it, but we must note the following facts.

In running through this cycle all three carbon atoms of the pyruvic acid molecule are oxidized to carbon dioxide at the expense of the oxygen of water, while hydrogen is constantly being led away from the cycle by means of NAD and the corresponding enzymes − oxidoreductases. Direct splitting off of carbon dioxide is accomplished by carboxylases containing thiamine pyrophosphate (TPP).

Thus we see that the same categories of enzymic mechanisms as in anerobic metabolism are active here, but the order of the reactions has been significantly changed. The main difference consists in the fact that the released hydrogen does not go to waste, but is used to get considerable additional quantities of energy through its oxidation by atmospheric oxygen. The intermediate products formed in the cycle join it with other metabolic systems, as a result of which a direct connection is established as well as a mutual dependence between the metabolism of carbohydrates, fats, organic acids, and proteins. For example, the keto acids formed in the cycle are converted by direct interaction with ammonia, that is, by direct ammoniation or through a transamination reaction, into alanine, aspartic, and glutamic acid; from them various

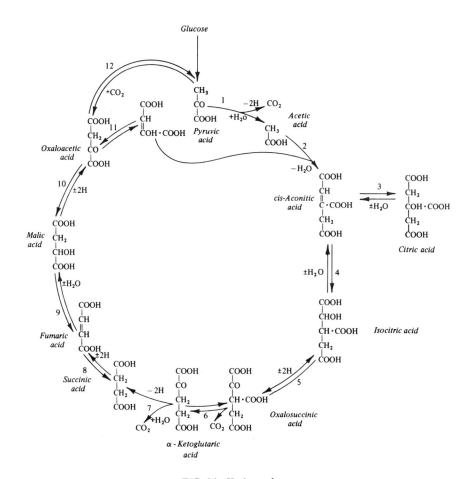

FIG. 31. Krebs cycle.

other amino acids are formed, which participate in the systems of proteins, hormones, enzymes, etc. (37).

The joining of the supplementary respiratory conversions to the chain of fermentation reactions can occur not only through pyruvic acid at the end of this chain, but also through its initial links.

In this case respiration is joined with the pentose phosphate cycle which in modern organisms exists mainly in the aerobic form.

The removal by NAD of hydrogen from intermediate links in the

Krebs cycle is not accompanied by the evolution of marked amounts of energy. Free energy of oxidation becomes available, not for the oxidation of the substrate but for the subsequent oxidation of NAD-H$_2$ by atmospheric oxygen; this does not occur in a single step but by way of the transfer of protons and electrons along the chain of a series of special oxidative enzymes. As a result of this transfer, each of the intermediate carriers, present in very small quantity, is constantly reduced at the expense of the substrate and is oxidized by molecular oxygen. Phosphorylation is usually coupled to this type of oxidation by oxygen, and is usually called oxidative phosphorylation (38). It is very significant, in essence the basic source of free energy for aerobic organisms. Actually it has been established at the present time that three pyrophosphate bonds (ATP) are formed by the oxidation of only one molecule of NAD, while during the fermentation of a whole molecule of sugar, only two such high energy bonds are formed.

The intermediate links in the oxidative chain in various organisms can be quite diverse substances (mediators) but the chief agents here are the flavoproteins As Mahler has shown (39), many different compounds of this type can be found in various representatives of the living world. In some of them the flavin group can be combined with nucleotides and other organic residues as well as with such metals as iron, molybdenum, or copper.

In some cases the flavin enzymes, obtaining hydrogen from reduced NAD, transfer it to porphyrin, a component of the cytochrome system, or to some other intermediate oxidizing mechanism which accomplishes the final oxidation of this hydrogen by atmospheric oxygen. In other cases flavoproteins which can receive hydrogen directly from the substrate and transfer it to the cytochrome system act as agents. Finally there are flavorproteins capable of transferring the hydrogen which they receive from some one of the systems directly to molecular oxygen.

The great diversity in the succession of reactions in the oxidative chain in different representatives of the animal and plant kingdoms indicates the relative youth of the system we are discussing, youthful in that it arose simultaneously in different organisms which were already at that stage of development in the living world when there was wide differentiation between separate parts of it.

We can arrive at the same conclusion on the basis of a knowledge of the different enzymes participating in the chain of oxidative transformations in various organisms. This particularly concerns the "terminal group" of those catalysts directly activating molecular oxygen (40). In organisms systematically far removed from each other, this problem is often solved by widely differing catalytic mechanisms. In addition to the very ancient flavin enzymes, we should cite first of all the cytochromes, which have been found in quite primitive aerobic organisms. With the appearance of molecular oxygen in the Earth's atmosphere, the cytochromes found in the most diverse living beings could easily begin to function as oxidase mechanisms, activating oxygen during respiration.

In connection with this the cytochromes and their corresponding enzymes — the cytochrome oxidases — are quite universal respiratory mechanisms. We find them in groups of organisms varying widely from a systematic viewpoint, but they have especially great significance in the respiratory process in many microorganisms as well as in the animal cell. In higher plants, a big role in this respect falls to the lot of the phenol oxidase system in which the enzymes are copper protein, and the hydrogen carrier is the "respiratory chromogen" of Palladin (41). These mechanisms are very specific for plants. Apparently they were added on during phylogenesis when the division of organisms into the animal and plant kingdoms occurred.

Peroxidase also has great significance in plant respiration; it activates the oxygen of hydrogen peroxide, while in the animal cell its role is comparatively small. Besides cytochrome oxidase, phenol oxidase, peroxidase and the flavin enzymes, the "final" oxidation by atmospheric oxygen can be catalyzed by ascorbic oxidase, lipid oxidase, and numerous other mechanisms.

In various living beings and in various stages of their life cycle the role of these individual mechanisms can vary within very wide limits. All this indicates the comparative phylogenetic youth of the process of respiration since it was added considerably later to the anaerobic process of energy metabolism.

We are led to the same conclusion by the great complexity of spatial configuration of protoplasm necessary for respiration. While fermentation and the anaerobic phosphorylation associated with it can be carried out in homogeneous solutions, the respiration mechanism and oxidative phosphorylation are firmly bound to specific structures of living matter. Attempts to accomplish these processes merely in a solution of the corresponding enzymes and mediators have been invariably unsuccessful.

It is evident that for successful transfer of protons and electrons along the chain of the oxidative systems, their accurate mutual spatial localization is necessary, otherwise the chain will be broken at one of its links. Particularly sensitive in this respect is the bond between oxidation proper and phosphorylation. Thus, for example, if certain concentrations of specific inhibitors are used, "respiration" can still be preserved, but it is irreparably "uncoupled" from phosphorylation.

The most primitive structural formation which we can find even in the bubbles of Goldacre or on the surface of the coacervate drops of Bungenberg de Jung is the protein-lipid membrane. A similar membrane can be also found on the surface of many bacteria. If they undergo lysis, these membranes can be isolated in the form of so-called "ghosts." The investigations of Gel'man (42) of quite primitive bacteria, but those still capable of respiration, have shown that in these bacteria the enzymes of the electron transport chain are incorporated in the surface of protein-lipid membranes (Fig. 32). The respiration process as a whole can be accomplished only when the spatial "montage" of enzymes is preserved undisturbed.

In modern higher organisms the spatial configuration of respiration has attained still more complexity and perfection. Here the respiratory function is accomplished by structural formations specifically adapted for this purpose – the mitochondria (Fig. 33) (43). Their very fine internal structure, found only by electron microscopic investigation, is basically characterized by the presence of a very well-developed system of protein-lipid membranes, to a certain degree analogous to those which we noted above for chlorophyll-containing granules. Thus the system of protein-lipid membranes can be regarded as a very general principle of spatial configuration of living matter.

Owing to the strictly specific interarrangement of enzyme complexes in an ultrafine structure, the mitochondria have attained an exceptional functional efficiency of the whole apparatus. In particular, for example, the high efficiency of electron transfer in the flavin proteins and hemin enzymes depends, according to Green (44), on the fact that the contact between these groups of enzymes is maintained here in the mitochondria by a lipoprotein membrane.

Thus the mitochondria of modern higher organisms, like the chloroplasts of the plants, are very complex and efficient structures, well adapted to the fulfillment of specific biological functions.

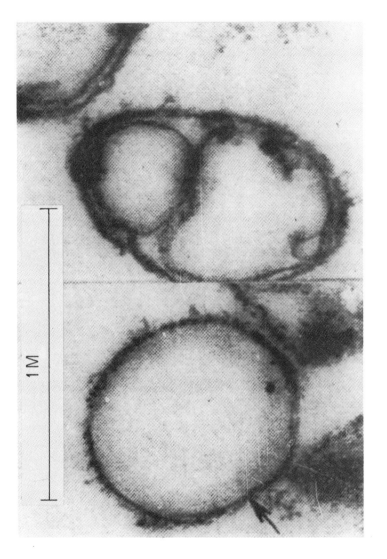

FIG. 32. Protein-lipid bacterial membrane of *Micrococcus lysodeikticus.*

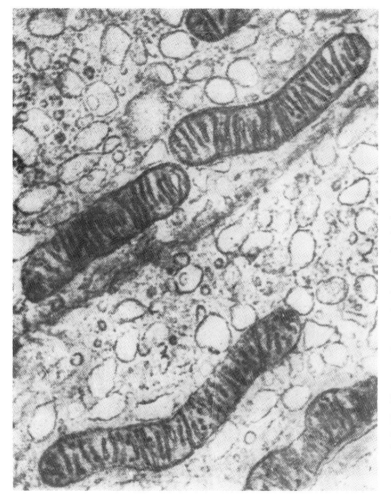

FIG. 33. Electron microscopic photograph of mitochondria of mouse epidermis.

It is clear that such structures could have arisen only during a very long development of living beings, on the pathway of improvement of their aerobic metabolism, considerably later than the large-scale formation of oxygen in the Earth's atmosphere. The circumstance that the structure of the mitochondria can exist only at a relatively high partial pressure of O_2 is particularly indicative of this. As the investigations of Gavodan and his colleagues have shown, mitochondria vacuolize and break down if the content of oxygen in the surrounding medium becomes less than 30% of its usual level.

It is too bad that we do not yet have data which would permit us to represent the pathway of step-wide formation of mitochondria from more primitive protoplasmic structures during the evolutionary development of organisms. Yet in large measure this can be said about that other very important cellular structure, the cell nucleus.

Of course, the intramolecular structure of DNA, formed at some time on the basis of the division and increase in functional efficiency of nucleic acids, must have been evolving for a long time during the development of life. This structure became better and better adapted to solution of the problem of very accurate self-reproduction and transfer of genetic information. However, a very important role in this respect must have been played also by the development of those multimolecular structures which DNA was principally responsible for forming. This was particularly important for solution of the problem of uniform distribution of DNA during cell division.

In bacteria and blue-green algae we find spatial configuration still at a relatively low level of development. Here, unlike more highly organized cells, the nuclear material is simply located in the center of the protoplast in the form of spherical or twisted shapes, consisting of DNA and possessing chemical properties characteristic of the nucleus. However, these formations do not correspond in their structure to typical cell nuclei of higher organisms, since they do not have the characteristic internal structure differentiation and are not set apart by a distinct membrane from the cytoplasm surrounding them. When the cells of bacteria and blue-green algae divide, a simple division of the nuclear material into two daughter fragments probably occurs.

The role of the internal nuclear structure increases radically in significance in relation to the phenomenon of cell copulation and in particular in relation to the reproductive process. After the latter arose, the

problem of proper division of nuclear material was immeasurably complicated. This led to the formation of the nucleus – a new organization,
astounding in its delicacy and accuracy, and capable of mitosis. Of
course, this organization could have been added only during a further
very long evolution of a quite high degree of development.

Thus the evolutionary development of cells in their modern form, the
cells which we usually consider as the most primary indivisible elements
of life, required enormous interludes of time, changes in numberless
generations of precellular living beings. Some modern authors (45) resur-
recting the old theory of Merezhkovskii (46) on symbiogenesis, even
consider it possible that in the initial phase of the existence of life,
individual structural formations developed as independent protobionts or
primitive organisms and only then were united in that most complex of
biological complexes which is the cell.

The formation of organizations peculiar to all modern living beings
both in space and in time, the formation of foundations of biological
metabolism and cellular structure can be understood only by a study of
the history of evolution of life and the establishment of biological
relationships specific for this evolution.

The formation of these foundations during the development of life
required many many hundreds of millions of years, perhaps half of all
the time during which life has existed on the Earth.

Thus, attempts to reproduce directly and artificially, to synthesize
even the most primitive modern living being even now still seem very
naive. Apparently the synthesis of life must be begun with the same
systems which were the starting point of our life on Earth.

From the metabolism of living bodies and the ultrafine structures
peculiar to them, a series of properties are directly derived which are
required for any living being known to us now, properties which in their
aggregate qualitatively distinguish organisms from objects of the inorganic world. These are the ability of living bodies to absorb actively and
selectively material from the surrounding medium, and in reverse to
excrete products of metabolism into this medium; further, there is the
ability to grow, multiply, reproduce, move about in space, and finally
there is the reciprocal reaction to external stimulation characteristic for
all life – irritability.

Each of these properties in the process of further development of
organisms not only became more and more complex, but also was

converted into qualitatively new forms of the manifestation of life.

Since up to the present the evolution of life has flowed not along a single channel, but has been developed along numerous branching pathways, these new manifestations were inherent not for the whole living world, but only for some particular part of it. However, we must not ignore them, if we want to obtain a realistically exhaustive representation of life.

The further the evolution of living beings proceeded, the more these new manifestations of life became increasingly complex and dependent on biological consequences. Thus they could not be mechanically reduced to an elementary process of inorganic nature, eluding the evolutionary pathway of development of living matter. Their actual understanding is attainable only by studying the history of their emergence from more primitive forms of biological organization.

REFERENCES

1. E. Zuckerkandl and L. Pauling. *In* "Problemy evolyutsion. i tekhnicheskoi biokhimii." ("Problems of Evolutionary and Technical Biochemistry"), p. 54. Izd. "Nauka," Moscow, 1964.
2. A. Oparin. *Proc. 5th Intern. Congr. Biochem., Moscow, 1961* Vol. 3. Pergamon Press, Oxford, 1963; S. Fox, ed. "The Evolution of Life," Vol. 1. Evolution after Darwin. Univ. of Chicago Press, Chicago, Illinois, 1960.
3. M. Florkin. "Biochemical Evolution." Academic Press, New York, 1949. *In* "Comparative Biochemistry" (M. Florkin and H.S. Mason, eds.). Academic Press, New York, 1949.
4. S. Cohen. "On Biochemical Variability and Innovation." Library Marine Biol. Lab., Woods Hole, Massachusetts, 1962; *Advan. Virus Res.* 3, (1955).
5. C. Ternetz. *Jahrb. Wiss. Botan.* 51, 435 (1912).
6. A. Artari. "K voprosy o vliyanii sredy na formu i razvitie vodoroslei" ("Effect of the Medium on the Form and Development of Algae"). Moscow, 1930; R. Harder. *Z. Botan.* 9, 161 (1930); S. Goryunova. "Khimicheskii sostav i prezhiznennye vydeleniya sine-zelenykh vodoroslei" ("Chemical Composition and Vital Secretions of Blue Green Algae"). Izd. Akad. Nauk SSSR, Moscow, 1950.
7. E. Aubel. *In* "Origin of Life on Earth" (A. Oparin, ed.). Pergamon Press, Oxford, 1959.
8. G. Ehrensvard. *Ann. Rev. Biochem.* 24, 275 (1955).
9. V. Tauson. *Priroda* 6, 43 (1934); "Osnovnye polozheniya restitel'noi bioenergetiki" ("Fundamentals of Plant Bioenergetics"). Izd. Akad. Nauk SSSR, Moscow, 1950; C. Zobell. *Advan. Enzymol.* 10, 433 (1950).
10. A. Thaysen. *Proc. 3rd Intern. Congr. Microbiol., New York, 1939,* p. 729.

Waverly Press, Baltimore, Maryland, 1939; W. Rosenfeld. *J. Bacteriol.* **54**, 664 (1947).
11. H. Krebs and H. Kornberg. "Energy Transformations in Living Matter." Springer, Berlin, 1957.
12. A. Terenin. "Fotokhimiya krasitelei" ("Photochemistry of Dyes"). Izd. Akad. Nauk SSSR, Moscow, 1947.
13. A. Krasnovskii and A. Umrikhina. *Dokl. Akad. Nauk SSSR* **155**, 691 (1964).
14. A. Szutka. *In* "The Origin of Prebiological Systems" (S.W. Fox, ed.). Academic Press, New York, 1965.
15. M. Dolin. *Bacteria* **2**, 425 (1961).
16. A. Krasnovskii. *Dokl. Akad. Nauk SSSR* **60**, 421 (1948); **103**, 283 (1955).
17. A. Krasnovskii. *Zh. Fiz. Khim.* **320**, 968 (1956).
18. A. Krasnovskii. *Izd. Akad. Nauk SSSR, Ser. Biol.* **2**, 122 (1955).
19. E. Kondrat'eva. "Fotosinteziruyushchie bakterii" ("Photosynthesizing Bacteria"). Izd. Akad. Nauk SSSR, Moscow, 1963.
20. F. Muller. *Arch. Microbiol.* **4**, 131 (1933).
21. I. Foster. *J. Bacteriol.* **47**, 355 (1944); *In* "Bacterial Physiology" (C. Werkman and P. Wilson, eds.), p.361. Academic Press, New York, 1951.
22. C. van Niel. *Arch. Microbiol.* **3**, 1 (1931); **7**, 323 (1936).
23. C. van Niel. *In* "Photosynthesis in Plants" (Y. Franck and W. Loomis, eds.), p. 437, 1949.
24. D. Arnon. *Proc. 5th Intern. Congr. Biochem., Moscow, 1961* Vol. 6. Pergamon Press, Oxford, 1963.
25. H. Gaffron. *Am. J. Botany* **27**, 273 (1940); *Plant Physiol.* **1A** (1960).
26. H. Gaffron. *In* "Horizons in Biochemistry" (M. Kasha and B. Pullman, eds.). Academic Press, New York, 1962.
27. E. Rabinowitch. "Photosynthesis and Related Processes." Wiley (Interscience), 1945 – 1956.
28. M. Calvin. *Proc. 3rd Intern. Congr. Biochem., Brussels, 1955* p. 211. Academic Press, New York, 1956.
29. A. Oparin, El Kharat'yan, and N. Gel'man. *Dokl. Akad. Nauk SSSR* **157**, 207 (1964).
30. W. MacElroy and H. Seliger. *Proc. 5th Intern. Congr. Biochem., Moscow, 1961* Vol. 3, p. 159. Pergamon Press, Oxford, 1963.
31. B. Isachenko. *Mikrobiologiya* **4**, 964 (1937).
32. S. Kuznetsov. *Mikrobiologiya* **17**, 307 (1948).
33. Y. Sorokin. *Proc 5th Intern. Congr. Biochem., Moscow, 1961* Vol. 3, p. 171. Pergamon Press, Oxford, 1963.
34. B. Horecker. *Proc. 5th Intern. Congr. Biochem., Moscow, 1961* Vol. 3, p. 86. Pergamon Press, Oxford, 1963.
35. K. Bernahauer. *Ergeb. Enzymforsch.* **3**, 185 (1934).
36. H. Krebs. *Proc. 2nd Intern. Congr. Biochem., Paris, 1952* p. 42. Masson, Paris, 1953.
37. A. Braunshtein. "12–3 Bakhovskoe chtenie." Izd. Akad. Nauk SSSR, Moscow, 1957.

38. A. Kotel'nikova. *In* "Problemy evolyutsion. i tekhnich. biokhimii." ("Problems of Evolutionary and Technical Biochemistry"), p. 80. Izd. "Nauka," Moscow, 1964; V. Skulachev. *Usp. Biol. Khim.* **6**, 180 (1964).
39. G. Mahler. *Proc. 3rd Intern. Congr. Biochem., Brussels, 1955* p. 264. Academic Press, New York, 1956.
40. D. Mikhlin. "Biokhimiya kletochnogo kykhaniya" ("Biochemistry of Cell Respiration"). Izd. Akad. Nauk SSSR, Moscow, 1960.
41. A. Oparin. *Biochem. Z.* **282**, 155 (1927).
42. H. Gel'man. *Usp. Sovrem Biol.* **47**, 152 (1959); *In* "Problemy evolyutsion. i tekhnich. biokhimii" ("Problems of Evolutionary and Technical Biochemistry"), p. 159. Izd. Akad. Nauk SSSR, Moscow, 1964.
43. A. Lehninger. "The Mitochondrion." Benjamin, New York, 1964.
44. D. Green. *Proc. 5th Intern. Congr. Biochem., Moscow, 1961* Plenary report. Pergamon Press, Oxford, 1963; *In* "Horizons in Biochemistry" (M. Kasha and B. Pullman, eds.). Academic Press, New York, 1962.
45. H. Ris and W. Plant. *J. Cell Biol.* **13**, 383 (1962).
46. S. Merezhkovskii. *Biol. Zentr.* **320**, 278, 321, and 353 (1910).

CONCLUSION

In conclusion we will try to give some objective chronology to those events described in previous chapters, since without this we cannot regard the history of the origin of life as completely finished. However, at the present time, we can only discuss (and that quite hypothetically) the sequence of individual stages of biogenesis; absolute dates established by paleontological findings belong mainly to very much later events in the development of life on Earth.

If we begin to compute the age of our planet at that period when it was beginning to be formed from the material of the proto-planetary cloud, we can assume this age to be approximately 5 billion years in round figures. So if we wanted to represent the whole history of the Earth we would have to write 10 volumes of 500 pages to a volume, each page of which would correspond to a period of 1 million years. Until recently we considered ourselves able to read in sequence only one volume, the tenth and last of this history.

The fossilized remains of plants and animals preserved since the beginning of the Cambrian period permit us to represent clearly the irreversible process, continually evolving over the course of 500 million years, of the sequential development of the organic world. The very important individual stages succeeding each other in this long route in the development of life are so well represented in the paleontological chronicle of the Earth of this period that the whole process can serve as a dependable chronometer by which geologists determine the time of formation of the sedimentary rocks they are investigating.

Turning over the pages of this volume one by one, studying the chronology of these important events which so enriched the history of the development of the organic world during the last 500 million years, we are immediately made aware of the steady increase in pace of this development. At the beginning of this period this pace was relatively slow, considerably slower than at its end. For example, almost half of this period was required for plants completely to conquer dry land, and at the edges of the seas and marshes amphibious animals crept out, still remaining completely dependent on the water.

187

Later developments in the terrestrial animal world went much faster, but all the same it took about 100 million years before the supremacy of reptiles was established on the Earth; they reached their apogee about 60-70 million years ago. Only half of this time, the next 35 million years, elapsed before the reptile kingdom gave way to the kingdom of birds and animals, which acquired the structural traits completely familiar to us now only about 5-7 million years ago; only a single final page of our volume embraces the whole history of mankind.

The high and ever-increasing rate of the development of life during this period also was responsible for the astonishing diversity of highly organized plants and animals in the world surrounding us. But the beginning of life, of course, did not in any way correspond with the beginning of our tenth volume. On the contrary, on the very first pages we meet with a quite luxuriant development of life, with a great many species of multicellular algae and with a variety of invertebrate animals: jellyfish, worms, echinoderms, mollusks and trilobites. Only vertebrates are lacking. However the plants and animals we have enumerated are already comparatively highly organized living beings which came to emerge only as a result of very long development of life preceding the Cambrian period.

Until comparatively recently it was considered that the paleontological record stopped with the Cambrian period; that the remains of organisms inhabiting the Earth more than a half billion years ago had not survived to the present, since the rock formations incorporating them had been subjected to such widespread changes and metamorphosis, completely obliterating all biological structures. However, recently it has been successfully established that at many points of the Earth, particularly on such platforms as the Russian, Siberian, or Chinese, beneath layers with the oldest complexes of Cambrian fossils are found clearly related strata of very slightly altered rocks, sometimes of quite considerable thickness. They underlie the Cambrian deposits and compose, with them, a single whole.

These strata have received different names in different parts of the world (Sinian, Eocambrian, Belt, etc.). In the USSR, as suggested by Shatskii, they are set apart in the so-called Riphean group, belonging to the Proterozoic Era. The study of these rocks began comparatively recently, but already many fossil formations preserved from the Precambrian organic world have been found in them (1). These investigations indicate that late Precambrian life was a direct predecessor of the Cambrian. The fossils found here demonstrate the existence at that time of a quite abundant flora and fauna, already comparatively highly developed, not only of unicellular but also multicellular organisms. Radiolaria, Foraminifera, and siliceous sponges for example, have been found in carbonaceous Precambrian Bretonian shales. Wolcott found Polychaeta in Precambrian formations of the Grand Canyon in North America. In Sweden, imprints of an articulated animal have been found, perhaps ancestors of the Cambrian trilobites. Vologdin (2) has found spines of siliceous sponges, very primitive in structure, in the Precambrian of the USSR and China, imprints of worms later developing to a greater extent in the Cambrian period, and other traces of animal life of the late Proterozoic Era.

However, there are no animal remains in deeper layers, perhaps because the primitive animals of this period did not have the ability to form skeletons or shells, and their soft tissue disappeared without a trace during these immense intervals of time which separate us from the Precambrian period.

In contrast to this, the plant world is represented in these rocks by numerous fragments and aggregations of various algae and spores of some sort of primitive plants (Fig. 34) (3). We can glimpse in them the epoch of life preceding the Cambrian period — the Proterozoic Era. This era is a tremendous stage in geological development, beginning almost 2 billion years ago; that is, in that period extremely important to life on the Earth's surface when the atmosphere of our planet began to lose its reducing character, gradually being enriched with free oxygen.

Usually the whole of this enormous era, surpassing in its extent more than twice the entire succeeding era, is divided into three geological, paleontological systems:
1. Sinian (beginning about 1200 million years ago and lasting up to the Cambrian period).
2. Eniseian (beginning 1500 and ending 1200 million years ago).
3. Sayanian (beginning 1900 and ending 1500 million years ago).

Thus the Sinian period alone encompasses almost all of the ninth and the second half of the eighth volume of our history of the Earth, the Eniseian system occupies the beginning of this volume, while the Sayanian period takes up the greater part of the seventh volume.

Unfortunately many pages of these volumes have been quite spoiled

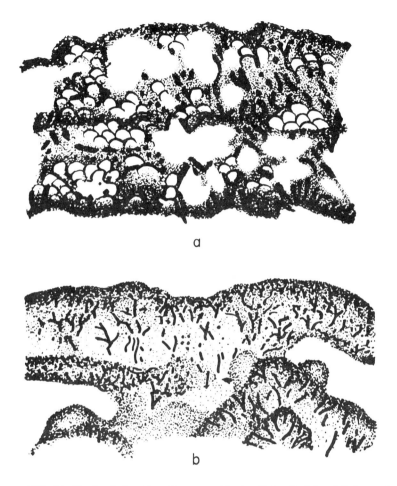

FIG. 34. The fossil algae *Somellostroma visiculare* Vologdin (a) and *Pralarenaria bullulata* Vologdin (b). Lower Cambrian.

by time, many of them entirely wiped out, the sequence of others greatly confused and distorted. In particular, in several cases we have had to recognize that the layers of Riphean formations sometimes contain plant fragments and spores of a considerably younger age than would be expected from the geological data of that location. Apparently they were sometimes carried there from overlying strata. We must

FIG. 35. Typical construction of stromatolites.

therefore approach with special care such accidental nonsystematic findings in strata in Riphean and even more ancient rocks.

Nevertheless we can now form some idea of the development of life in the Proterozoic Era, although a very general and approximate one. We will not find here the brilliant variety of modern organic forms and those rapid changes with time which are characteristic of our tenth volume of the history of the Earth. We can imagine that the evolution of life at this stage went at a less vigorous pace than in the succeeding Cambrian period.

Basically the Proterozoic Era was one of algae and bacteria (4). As we have seen, the most primitive representatives of multicellular animals appeared only towards its end. Of course we can imagine that unicellular animals, protozoa in particular, existed at an earlier period, but we have almost no factual material concerning this. In contrast, the plant world of the Proterozoic Era left behind it numerous and completely trustworthy monuments.

Chief among them are the so-called stromatolites (5), peculiar limestone formations constructed in the form of relatively regular cupolas, consisting, as it were, of a series of cones set inside each other (Fig. 35). They attain up to a meter in height and a half-meter in cross-section at the base. Sometimes they are large bodies with a rounded surface, sometimes they are made up of individual incrustations. In other cases the stromatolites have a conical or even a pointed form. This form is characteristic both of the Precambrian period and for the beginning systems of the Paleozoic. Clumps of stromatolites have formed reefs and

FIG. 36. Form of stromatolite ravine, formed by the alga *Crustophycus ongaricus*.

extensive stromatolite ravines, characteristic of shallow waters of the early Sinian seas (Fig. 36).

Broad laboratory investigations of stromatolites, particularly the study of sections made from them, have shown without any doubt that these formations are products of the life activity of ancient photosynthesizing organisms. A large number of species have even been successfully established and described, mainly of blue-green, but also red algae, which later participated in the formation of stromatolites (Fig. 37).

In general both marine and freshwater algae were apparently well represented in the various periods of the Proterozoic Era, particularly in the Sinian. The upper complex of the Riphean formations, specifically, including the corresponding series in the Urals, the Russian platform and the upper Sinian of the Chinese platform are characterized by vigorous development of the branching alga *Collenia*.

In the lower complexes, the algae-stromatolite formers should be mentioned first; in a number of cases they are rock-forming. For example, the lowest Riphean beds of the Urals and the lowest Sinian beds of the Chinese platform are characterized by such species as *Canophyton cylinducus*.

The formations found in the Sahara (6) of limestone secreted by organisms are classified as the first appearance of stromatolites. These

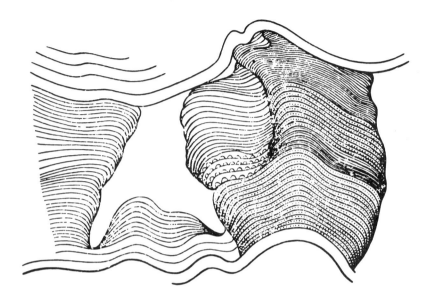

FIG. 37. Types of stromatolites formed by various algae.

algal reefs are apparently the earliest bioherms* known to geological history. They are approximately 1 billion years old.

However, Vologdin indicates that the majority of the oldest algal species did not form discrete bodies, unchanging in form, of the stromatolite type, but were stratified in sediment, converting it into dense rock, where the morphology of colonies and products of their life activity were only partially preserved. Very few studies have as yet been made of these oldest algae. However, in marine deposits, formed in the Sinian period, this author has recently discovered a wide variety of genera and species of blue-green and red algae (7).

This period can be characterized as the second half of the transitional period between the pre-actualistic and actualistic epochs on the Earth's surface. At this time a considerable enrichment of the atmosphere with free oxygen had already occurred and it can be imagined that the early stromatolites played no small part in this enrichment. Therefore they must already have had photosynthetic ability. However, we know that

* Bioherms are geological formations of biogenic origin.

photosynthesizing organisms must have had a very highly differentiated and very efficient intracellular apparatus, which could have been formed only as the result of very long evolution of the first living beings, only very much later than the emergence of life. In the exploration of life, therefore, we must delve deep into a still older period in the existence of our planet. But the traces of life preserved from these periods are extremely scanty and are always random in character. This detracts strongly from the reliability of the conclusions reached on the basis of their study.

Vologdin has made the very interesting assertion that the blue-green algae in the stromatolites must often have been accompanied by iron bacteria, sometimes predominating in the periods between vegetative seasons and forming iron films in the stromatolites. The iron bacteria (8), being chemosynthetic and obtaining their energy by oxidizing ferrous oxide to ferric, have an unconditional requirement for molecular oxygen, which in the transitional epoch we are talking about would have been present in the atmosphere in considerably lesser amounts than now. They could have obtained part of it from algae, with which they were symbiotic, and together caused a stromatolite concentration of ferric hydroxide and calcium carbonate. "Sometimes," writes Vologdin (9), "the activity of the bacteria alternated with the activity of the algae: summer was the time for optimal development of the algae, winter the time of predominant development of the iron bacteria." As a result of this, a complex of certain stromatolites was created, where limestone alternated with iron hydroxide (Fig. 38). Vologdin considers that the iron bacteria already existed at the very beginning of the Proterozoic Era and even in the beginning of the Archean Era. However, in order to confirm this it would be necessary to have much more factual material than we now possess.

The very oldest authentic fossils are now considered to be the fragments of organisms described by Tyler and Barghoorn (10) preserved in the iron ores of Southern Ontario (Canada). These organic fragments, under the conditions of their habitat, were probably converted into fossils by substitution of their initial substance by silicon, molecule for molecule. In this way their internal structure was preserved in the finest detail. On the basis of a study of this structure, the conclusion was reached that some of the forms found are primitive algae and other fungi.

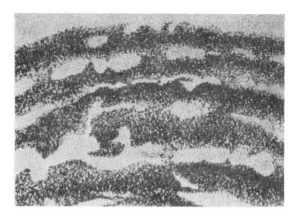

FIG. 38. Layer of algal deposits forming stromatolites.

It is apparent that such a conclusion is very tentative and preliminary and on the basis of it, of course, it is completely impossible to solve the problem of the nature of the metabolism in these organisms. For example, it is impossible to say whether these forms, which belong to the "algae" actually have the ability to photosynthesize. It is also very difficult to determine the absolute age of these fossils. Now, on the basis of rubidium-strontium dating, their age has been conditionally established at 1600 million years. Therefore, the organisms represented by the fragments found in Ontario lived in the transitional epoch and could have been either anaerobes or aerobes.

We can be completely certain, however, from the comparative biochemical data which we have presented, that the birth of life and the beginning steps in its development occurred in the reducing atmosphere, in the pre-actualistic epoch. Geologists date the end of this epoch at 2 billion years ago, and thus the emergence of life must be relegated to an even more remote period.

The earliest manifestations of life now are considered to be the limestone secretions discovered by McGregor (11) in the dolomite series in Southern Rhodesia (South Africa). These do not include authentic fossils with structurally preserved remains of early organisms. But in the limestone a layered structure of cones set inside each other is very distinctly seen, which it would be difficult to interpret as the result of abiogenic processes. To a certain degree these formations are similar in their structure to deposits formed by the action of lime-secreting algae (stromatolites), which, however, inhabited a very much later geological epoch. Because of this similarity the Rhodesian deposits are often called algal limestones. But of course, this is incorrect since it is impossible to draw conclusions regarding the definite origin of the organisms forming these deposits from only a single structural similarity in the secreted deposits.

The absolute age of the limestone secretions of Rhodesia has been assumed to be approximately 2700 million years on the basis of isotope analysis of the granite veins cutting through them. They undoubtedly belong, then, to the pre-actualistic epoch, and the tendency to ascribe to the organisms forming them the same metabolism as in the much later aerobic algae on the basis of only a single form of deposit is completely groundless; the more so since lime secretions can easily occur in anaerobic metabolism. In addition to this, there is some basis for assuming that the producers of Rhodesian lime secretions did not have a definite morphological structure but were irregular masses. It may be that they were not even primitive organisms, but our hypothetical protobionts, and in that case, we have closely approached the period of the initial genesis of life.

Investigation of still earlier rock formations has not yet revealed any direct signs of life, and we can form our conclusions regarding these remote periods of the Earth's history only on the basis of a study of geochemical materials (12). However, results thus obtained often can be interpreted in different ways. For example, quite valuable data can be obtained by isotopic analysis of graphite deposits or Precambrian coal. But for the oldest of them we still cannot say whether these beds were formed from the organic matter of the "primitive soup," of coacervate drops, protobionts, or from living matter already formed.

In order to view at a single glance the entire pathway of evolution that we have passed over, an approximate outline is presented here in Fig. 39. Time, expressed in billions of years, is plotted on the horizontal base line; the time which separates us from some particular event. Above this line, the basic boundaries of the evolutionary stages of our planet are marked off (of course, very approximately), the stages in the evolution of carbon compounds that appear on the route to the origin and development of life are given below the line.

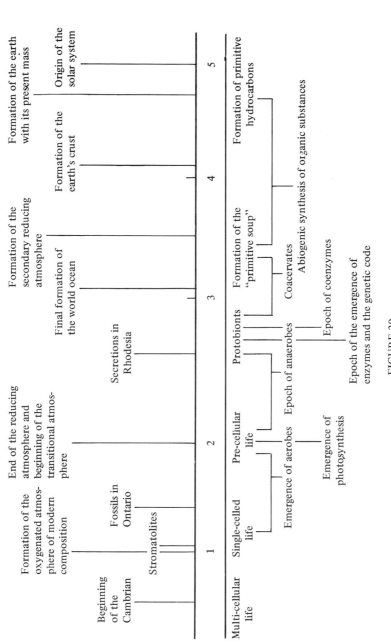

FIGURE 39

Examination of this outline permits us to realize the whole enormity of our ignorance and the whole extent of the horizons which will open before the biologists of the future. Now, on the basis of comparative morphological and paleontological data, we can quite completely visualize the course of biological evolution, beginning with the late Pre--Cambrian up to the present. But this is only a small area of our outline, located on the left. On the basis of astronomical, geological, and chemical data, we can, to a certain degree, visualize the first stages in the evolution of carbon compounds. Between these and the first section of the outline lies a long period of evolution, lasting a billion years, during which the basic principle changes in the organization of systems giving birth to living beings were slowly perfected. We can only hypothetically visualize the sequence, and to a certain degree the character, of these changes.

However, we still are unable to denote even an accurate time for the origin of life. That difference between the inorganic world and the world of living beings, which we can so easily establish now, arose only because all the intermediate forms of organization were eliminated long ago by natural selection. However, the evolutionary origin and subsequent development of life went through a series of intermediate steps; thus, the answer to the question we have posed is dependent on which of these stages we recognize as the beginning of life: the emergence of protobionts, already having an efficient metabolism, the emergence of proteins and the nucleic code, or, finally, the emergence of cells.

Scientific comprehension of the still very obscure periods of formation and organizational perfection of life first of all lies along the route of future vigorous development of evolutionary biochemistry, biophysics, cytology and physiology. Before Darwin, the science of biology accumulated an enormous amount of morphological material which acquired its scientific significance only when it was generalized into a single idea of development. Similarly, today, a vigorous accumulation of information is taking place, through the use of physical and chemical methods, on metabolic organization and on the structure of vitally important molecules, membranes, and organoids in presently living organisms, based on various degrees of evolutionary development. This material becomes exceptionally significant for the comprehension of the essence of life just at that point where it will be combined and perceived on an evolutionary basis.

Only by this type of evolutionary approach will we be in a position not only to understand which and how living bodies develop but also to answer the "hundred thousand ways" which inevitably arise on the pathway to a true comprehension of the essence of life; particularly to the question of why the whole organization of life from the molecular level to the level of the organism is so "purposeful," so well-adapted to constant self-preservation and self-reproduction in the given conditions of the surrounding environment.

REFERENCES

1. B. Keller. *Priroda,* No. 9, 30 (1959).
2. A. Vologdin. "Paleontologiya i poiski poleznkh iskopaemykh" ("Paleontology and the Search for Useful Fossils"). Izd. "Znanie," 1960.
3. S. Naumova. *Paleotologiya* 4 (1963).
4. A. Vologdin. *Izv. Akad. Nauk SSSR, Ser. Geol.* No. 2 (1947).
5. A. Vologdin. *Priroda* No. 9 (1955).
6. M. Gravelle and M. Lelubre. *Bull. Soc. Geol. France* [6] 7, 435 (1957).
7. A. Vologdin. "Drevneishie vodorosli SSSR" ("Oldest Algae of the USSR"). Izd. Akad. Nauk SSSR, Moscow, 1962.
8. N. Kholodnyi. "Zhelezobakterii" ("Iron Bacteria"). Izd. Akad. Nauk SSSR, Moscow, 1953.
9. A. Vologdin. "Zemlya i zhizn' " ("Earth and Life"). Izd. Akad. Nauk SSSR, Moscow, 1963.
10. S. Tyler and E. Barghoorn. *Science* 119, 606 (1957).
11. A. McGregor. *Trans. Geol. Soc. S. Africa* 43, 9 (1940).
12. P. Abelson. *Proc. 5th Intern. Congr. Biochem., Moscow, 1961* Vol. 3, p. 59. Pergamon Press, Oxford, 1963.
13. S. Manskaya and T. Drozdova. "Geokhimiya organicheskogo veshchestva" ("Geochemistry of Organic Substances"). Izd. "Nauka," Moscow, 1964.

Subject Index

A

Adenosine triphosphate, 144–145, 158–159, 169
  abiogeneic synthesis, 85
Aerobiosis, 174, 175
Albertus Magnus, 16
Algae, 155
Algal limestones, 196
Amino acids
  abiogenic formation, 76–79
  abiogenic polymerization, 83
Anaerobiosis, 156–157
  facultative, 175
Aquinas, Thomas, 17
Aristotle, 13, 16
  and spontaneous generation, 13
Asymmetry
  origin, 80–81
Atmosphere
  oxidizing, 174
  reducing, 55–57
  secondary
    see also Atmosphere, reducing
ATP
  see Adenosine triphosphate
Athiorhodaceae, metabolism, 166
Autotrophs, 167

B

Bacon, Francis, and spontaneous generation, 20
Basil the Great, and spontaneous generation, 15
"Bubbles" of Goldacre, 101–102

C

Cambrian period, 187–188
Carbon compounds
  see also hydrocarbons, nucleic acids, etc.
  abiogenesis, 34, 62, 64–65, 73, 91
  exogenous origin, 67
  extraterrestrial abiogenesis, 86–88, 90–91
Carbonaceous chondrites, 29, 60–62, 67
  composition, 86–87
Catalysts, 131–133
  see also Enzymes, Coenzymes
Cells, evolution, 183
Chemoautotrophs, 173
Chlorophyll, in coacervates, 123
Chloroplasts, 168–170
Coacervate drops, 105–124, 135, 146–147
Coenzymes, 133–135
Cometesimals, 52
Comets, composition, 59
Cytochrome system, 177–178

D

Democritus, 11–12
Deoxyribonucleic acid, 143–144, 146, 150, 182
  structure, 141
Deoxyribose, abiogenesis, 82
Descartes, Rene, and spontaneous generation, 20–21

200